N.R. Baker

Photosynthesis in relation to Plant Production
  in Terrestrial Environments
Natural Resources and the Environment Series
  Volume 18

Photosynthesis in Relation to Plant Production in Terrestrial Environments

by
C. L. Beadle, S. P. Long, S. K. Imbamba
D. O. Hall and R. J. Olembo

Published for the
UNITED NATIONS ENVIRONMENT PROGRAMME
by
TYCOOLY PUBLISHING LIMITED, OXFORD, ENGLAND

First published 1985 by
Tycooly Publishing Limited
England

© United Nations Environment Programme (UNEP)

ISBN 1-85148-003-3 (hardback)
ISBN 1-85148-004-8 (softcover)

All rights reserved. No part of the publication may be reproduced, stored in a retrieval system or transmitted in any form or by any means; electronic, electrostatic, magnetic tape, mechanical, photocopying, recording or otherwise, without permission in writing from the publisher.

Printed in Great Britain

CONTENTS

CONTENTS v
FOREWORD 1
AUTHOR'S PREFACE 3
LIST OF ABBREVIATIONS 5
SUMMARY 7

Part I PHOTOSYNTHESIS AND BIOMASS

Introduction 16
1. PHOTOSYNTHETIC MECHANISMS
   1.1 Photosynthesis as a process 17
   1.2 Light reactions 17
   1.3 Dark reactions 22
   1.4 Photorespiration 22
   1.5 The diffusion process 23
2. PHOTOSYNTHETIC DIVERSITY
   2.1 C_3 plants 27
   2.2 C_4 plants 27
   2.3 CAM plants 31
   2.4 "C_3/C_4 intermediates" 32
   2.5 Occurrence 33
3. PHOTOSYNTHETIC PRODUCTIVITY
   3.1 World Productivity 37
   3.2 Natural Ecosystems 40
   3.3 Agricultural Systems 47
   3.4 Biomass 50

Part II FACTORS INFLUENCING PHOTOSYNTHETIC PRODUCTIVITY

Introduction 54
4. LIGHT
   4.1 Available light 55
   4.2 Conversion of energy 57
   4.3 Quantum yield 59
   4.4 Light response curve 61
   4.5 Sun & shade species 62

5. TEMPERATURE
    5.1 Temperature 65
    5.2 Reversible effects 68
    5.3 Irreversible damage... 68
6. SOIL FACTORS
    6.1 Water 73
    6.2 Salinity.. 82
    6.3 Nitrogen and nutrients 85
7. POLLUTION
    7.1 Pollutants in the environment 91
    7.2 Gaseous pollutants.. 91
    7.3 Heavy metals 98
    7.4 Ultra-violet radiation... 99

Part III RESEARCH INTO PHOTOSYNTHETIC PRODUCTIVITY

Introduction 102
8. PHOTOCHEMISTRY AND CARBON METABOLISM
    8.1 Photochemistry, electron transport and photophosphorylation 103
    8.2 Carbon metabolism 106
9. GAS-EXCHANGE
    9.1 Controlled environments 109
    9.2 Natural environments 113
10. LIMITS TO BIOMASS PRODUCTION
    10.1 Light interception and growth 117
    10.2 Canopy structure.. 119
    10.3 Growth Analysis 122
    10.4 In the context of other factors limiting photosynthetic efficiency 124
11. INTEGRATIVE MODELS RELATING PHOTOSYNTHESIS TO
     PRODUCTIVITY.. 127
12. FUTURE PERSPECTIVES 131

REFERENCES 135

FOREWORD

Mostafa Kamal Tolba
Executive Director
United Nations Environment Programme

ALL OUR FOOD, most of our fuel and many of our fibres are derived directly or indirectly from photosynthesis. As the fossil fuel reserves, which are products of past photosynthesis, are depleted, current photosynthesis may become of even greater importance as a source of fuels and organic compounds. To some extent, this is already a reality as exemplified by the substitution of petroleum by alcohol derived from crops in Brazil and other countries. Since photosynthesis sets the ultimate limit on crop productivity, any improvement in photosynthetic efficiency can be seen as a means of increasing our ability to produce food, fibres and fuels. This may be crucial as the consumption of our reserves is at such a level that it may even be difficult to maintain the current capacity for producing both food and biomass particularly in developing countries.

The concept of current photosynthesis as the direct and major source of these commodities must be seen in the context of both social and environmental problems. At a social level competition between food, fuel, fibre or "chemical" crops for available land could create serious problems, especially in areas where food supplies barely meet demands, and where landowners might have the opportunity of replacing food crops by more remunerative fuel or other crops. At an environmental level, the improvement of photosynthetic efficiency, resulting from crop improvement and more extensive vegetation cover could alter the carbon cycle and affect atmospheric CO_2 concentrations. An alternative possibility is that natural vegetation of little direct economic value could be replaced by crops with a lower net photosynethic efficiency as has occurred with deforestation for the planting of agricultural crops and fuelwood. This would have the reverse effect of increasing the level of CO_2 in the atmosphere. Present predictions suggest that an increase in atmospheric CO_2 concentration will result in a corresponding increase in global atmospheric temperature and precipitation. In addition, concern is now being expressed about the continued atmospheric CO_2 enrichment which may favour certain plant species in preference to others.

In the field of agriculture, there has been an improvement in crop productivity resulting from an increase in the amount of total photosynthate invested into the portion of the plant of economic value, e.g. the grain of the wheat plant, the fibres of the cotton seed, and the trunk of the pine tree. High economic yields resulting from plant breeding research have largely been due to improvement of genotypes capable of producing large leaf area in response to fertilizer application. There has also been an upsurge of interest in research on gaseous pollutants and heavy metals on plant productivity. Current evidence suggests that present levels of some pollutants have deleterious effects on photosynthetic process and hence biomass production. There is, however, need to carry out further research in this field and particularly in view of the fact that gaseous pollutants know no territorial boundaries and that they also seem to have secondary effects, notably the effect on stratospheric ozone concentrations leading to an increase in ultra-violet radiation.

This book presents the current state of the art on the subject of photosynthesis by initially describing in part one the various photosynthetic mechanisms in relation to plant productivity. Part two of the book deals with the influence of environmental factors notably solar radiation, the weather, edaphic factors and pollutants on photosynthetic productivity. Part three concentrates on research which has hitherto been carried out on photosynthetic productivity at the chloroplast, leaf, whole plant and canopy levels and finally biomass production at field level. The authors are particularly well fitted to have written such a book, each one of them contributing in his own area of competence. This book should prove valuable to postgraduate students, researchers, agriculturists, environmentalists and all readers interested in photosynthesis in relation to bioproductivity.

AUTHORS' PREFACE

This book arises from a report entitled "Photosynthesis and bioproductivity" submitted for the 9th session of the governing council of UNEP (1981) and subsequently updated in 1984. It contains an extensive set of references. The introductory chapters (PART 1) introduce the subjects of photosynthesis and biomass production. The influence of environmental factors on biomass production through photosynthesis (PART II) and the relationship of different areas of photosynthesis research to improvement of biomass production (PART III) are covered in the remaining chapters.

The information presented considers the photosynthetic productivity of terrestrial plants only and it is inevitable that much of the contents relate to agricultural plants and agricultural systems since this is the area which has been most thoroughly researched. Every attempt has been made however to include relevant information from natural ecosystems and forests.

Photosynthesis has become widely researched in the last 20 years, and inevitably many books covering this topic have emerged. This book attempts to compress current knowledge of the process from chloroplast to whole plant level and particularly its relationship to the environment and for the production of dry matter. Problems of measuring productivity are also highlighted in practical and basic applications. The book is primarily aimed at those who require key information and key references in any area of terrestrial photosynthesis in a readily accessible form. It should also form a useful text for any potential research student in the field or the laboratory.

We would like to thank Mrs. Rita Bartlett of the University of Essex who provided invaluable assistance in the preparation of the manuscript.

Many people have made helpful comments on parts of drafts or have supplied useful information to us and in particular we wish to acknowledge help from the following: Dr. J. A. Berry, Dr. J. Coombs, Dr. J. E. Hällgren, Dr. M. M. Ludlow, Dr. S. Linder, Prof. J. L. Monteith and Prof. H. W. Woolhouse. The authors however take full responsibility for the final contents.

ABBREVIATIONS

$A, \bar{A}$	instantaneous or average leaf area ratio.
ATP	adenosine triphosphate.
ATPase	adenosine triphosphatase.
'B'	electron molecule transfer between Q and PQ.
$B_{max,min}$	maximum and minimum biomass of a plant stand or community over a 12 month period.
ΔB	Biomass change over a specified time interval.
$C, \bar{C}$	instantaneous and average crop growth rate.
C_a	CO_2 concentration of air.
C_c	CO_2 concentration at chloroplast.
C_s	CO_2 concentration at stomatal pore entrance.
C_w	CO_2 concentration at cell wall (intercellular concentration).
C_3	plants which initially fix CO_2 into glycerate 3-phosphate.
C_4	plants which initially fix CO_2 into oxaloacetate, malate or aspartate.
CAM	plants which initially fix CO_2 into malate at night and into glycerate 3-phosphate during the day.
CF_0	coupling factor 0 of ATPase/ATP synthase.
CF_1	coupling factor 1 of ATPase.
D	leaf area duration.
$E, \bar{E}$	instantaneous and average unit leaf rate.
F	flux of carbon dioxide (e.g. mg m^{-2} s^{-1})
G	direct plant losses to consumer organisms.
$L, \bar{L}$	instantaneous or average leaf area ratio.
L_d	plant losses by death and shedding.
LHCP	light harvesting chlorophyll-protein complex.
NADP/ NADPH-	nicotinamide adenosine diphosphate and reduced NADP.
PEP, PEPc	phosphoenolpyruvate, and PEP carboxylase.
PGA	glycerate 3-phosphate.
PQ, PQH$_2$	plastoquinone and reduced PQ.
P_G	gross primary production.
P_N	net primary production.
PS I	photosystem I.
PS II	photosystem II.
P700	reaction centre molecule of PSI.

P680	reaction centre molecule of PSII.
Q	Primary acceptor of P680 or quantum flux density.
r_a	boundary layer resistance.
r_s	stomatal resistance.
r_m	mesophyll resistance.
Σr	sum of resistances to CO_2 diffusion.
RPP	reductive pentose phosphate cycle.
R_t	total respiration.
RubP	ribulose *bis*phosphatase.
Ru*bis*CO	RubP carboxylase/oxygenase.
RWC	relative water content.
VPD	water vapour pressure deficit
Γ	CO_2 compensation point.
Ψ_p	turgor potential.
Ψ_π	osmotic potential.
Ψplant	plant water potential.
Ψsoil	soil water potential.

SUMMARY

1. Photosynthesis is a complex process which spans time scales from 10^{-13}s (light capture) to 10^7s (productivity). The light reactions of photosynthesis are located on the thykaloid membranes of the chloroplasts and generate assimilatory power in the form of ATP and NADPH. The separate stages of the light reactions are reasonably well understood, but with some notable exceptions.

2. Assimilatory power is utilized in the dark reactions of photosynthesis located in the stroma. These reactions hinge around the reductive pentose phosphate (RPP) or Calvin cycle, the only form of CO_2 assimilation which results in the net production of dry matter. A universal property of the carboxylating enzyme of the RPP-cycle (Ru*bis*CO) is an ability both to carboxylate and oxygenate its substrate ribulose *bis*phosphate (Ru*b*P). The oxygenation process causes a net loss of fixed carbon and is manifested in gaseous fluxes, termed photorespiration. Photorespiration, although decreasing the efficiency of light energy conversion, may be essential for the removal of excess photoreductant and possibly preventing photoinhibition.

3. CO_2 is present in low concentrations in our present atmosphere and may often be limiting to the photosynthetic rate. Its flux into the leaf and limitations to this flux have been summarized and analyzed by analogy to Ohm's Law, *viz* the concentration gradient between the atmosphere and sites of fixation with resistances in the boundary layer surrounding the leaf, at the stomata, and in the mesophyll.

4. Plants may be divided into distinct photosynthetic groups which cross some phylogenetic boundaries. They are then described as either C_3 or C_4 species depending on whether their first product of photosynthetic CO_2 assimilation is either a C_3 or C_4 compound. Two other photosynthetic groups are identified: Crassulacean acid metabolism (CAM) plants, and a tentative C_3/C_4 intermediate group to which only a few species have been ascribed.

5. C_3 species assimilate and reduce CO_2 entirely through the RPP-cycle. C_3 species occur in all habitats colonized by plants and are the most abundant photosynthetic type.

6. C_4 photosynthesis also incorporates the RPP cycle, but its distinctive biochemical feature is the concentration of carbon dioxide around chlorenchymatous bundle sheath cells, where the RPP-cycle is located, via the photosynthetic C_4 dicarboxylate cycle. This serves two functions: to increase the rate of photosynthesis and to suppress photorespiration under conditions of high light and temperature in which most C_4 species are found. C_4 plants can be divided into three categories, NADP-me type, NAD-me type, PEP-CK type, according to the

enzyme which catalyzes the decarboxylation of their C_4 products of carboxyalation.

7. CAM plants are succulents which may open their stomata at night to decrease daytime transpirational water loss and fix CO_2 into malate. Decarboxylation of malate occurs during the day and CO_2 enters the RPP-cycle while the stomata remain closed. This mechanism ensures net carbon fixation under conditions of severe water stress. In arid tropical climates the mechanism may decrease transpirational water loss per unit of carbon gained by as much as 95%.

8. The significance of C_3/C_4 intermediates is not yet clear. They are intermediate with respect to both the anatomical and biochemical features associated with C_3 and C_4 photosynthesis, and may conceivably form an evolutionary link between the two types.

9. The net annual primary production of the world (P_N), amounts to 8×10^{10} t (carbon) yr^{-1}. The energy content of this photosynthetic production considerably exceeds the world's annual energy use even though the solar conversion efficiency of the photosynthetic process is a mere 0.1%, averaged over the whole surface of the earth. The most important sources of this production are the natural forests and natural grasslands though current estimates of their productivity are based on measurements of minute and non-random samples. From the data available it is suggested that P_N may be underestimated in some instances by as much as 75% for above ground and even more for below ground production. Exudation of organic compounds from roots could also account for up to 50% P_N, a source which has been almost ignored. Year to year variation in P_N must not be overlooked, particularly in the semi-arid tropics and tundra where variation in rainfall and temperature, respectively, will have the most profound effects on production.

10. Detailed and precise estimates of productivity are available for agricultural systems but few studies include production of roots and rhizomes or parts of the plant which die and are shed before harvesting. Estimates of P_N vary between 1 and 88 t (dry matter) ha^{-1} and reflect the confounding of environment, cultivation practice and economic restraints on the production potential of crop plants. Maximum biomass production of C_4 plants exceeds that of C_3 plants, a consequence of the higher rates of photosynthesis of C_4 species and their ability to suppress photorespiration.

11. World food production currently exceeds that required by the world population but for each major cereal, production exceeds demand in the developed countries while the opposite pertains to developing countries. Substantial increases in production could be realized by increasing the productivity of existing cultivated land.

12. Biomass production as a source of fuel for cooking and heating and as an alternative and renewable source of energy is receiving increasing attention. The major requirements are for quantity and energy value and a net energy yield i.e. the energy content of the biomass produced must exceed the energy put into the system to harvest and produce the biomass product.

13. Light intercepted by chlorophyll is the discriminant of biomass production. It can be shown that the maximum conversion efficiency of solar radiation into photosynthetic products is 3.7–4.4% and 5.0–5.8% in C_3 and C_4 plants

respectively. Annual conversion efficiencies are considerably less due to incomplete light interception, agricultural practice, pests and diseases, genetic limitations, growth patterns, assimilate partitioning and harvest yield.

14. The quantum yields (φ) of C_3 and C_4 plants are similar at 30°C, though in C_3 plants there are marked changes in φ with temperature because of photorespiration, whilst in C_4 plants φ remains constant with temperature. The remarkable consistency of φ within each photosynthetic type suggests that it may be a relatively conservative property of green plants and not therefore amenable to easy manipulation.

15. Individual leaves of most C_3 plants are unable to use additional light above 500 µmol m^{-2} s^{-1} (roughly 25% full sunlight) but this is not true of C_4 plants which fail to saturate even at full sunlight. Photosynthetic capacity is a function of the environmental conditions to which the plant is subjected during its growth and species integrate and adjust several partial photosynthetic processes to the available quantum flux density.

16. Temperature is often the most important factor determining biomass production. The shape of the response curve of photosynthesis is plotted against temperature and the ability of plants to adapt to changing temperature is species dependent. Adaptation to the prevailing temperature conditions has a role in plant survival.

17. Reversible effects of supraoptimal temperatures on photosynthesis are partly explained by the increase in photorespiration in C_3 plants and partly to the decline in the rate of supply of RubP linked to a marked decline in coupled electron transport in C_3 and C_4 plants. At suboptimal temperatures it is a function of the rate limiting dark reactions.

18. The irreversible effects of supraoptimal temperature stress on photosynthesis are related to the lipid properties of the membranes which are heat labile. There is a breakdown of energy transfer between the light harvesting molecules, the reaction centre and the electron transport chain. At suboptimal temperatures, these irreversible effects are partially caused by reduced stomatal conductance and partially by phase separation of the chloroplast membrane which disrupts electron transport. Low temperature inactivation of some rate limiting enzymes of the dark reactions may also occur.

19. The availability and utilization of water and mineral nutrients are major factors, which either modify the response of biomass production to temperature, even in temperature and humid climates, or are themselves the dominating influence on production.

20. Although the effects of water stress on productivity are in some circumstances offset by mobilization of storage compounds, the effect of water stress on current photosynthesis also contributes to the loss of yield. This is manifested partly by stomatal closure and partly by inhibition of photosynthetic processes which are non-stomatal in origin. These include all facets of the photosynthetic process as well as an increase in the ratio between photorespiration and photosynthesis in C_3 plants, a change which may decrease photoinhibition of photosynthesis by recycling CO_2 under the prevailing conditions of low

inter-cellular CO_2 concentrations in water-stressed plants.

21. There is considerable evidence that biomass production is correlated with water use. The major factor lowering production under water stress appears to be reduced light interception as a result of decreased leaf areas. The efficiency of water use by C_4 plants is higher than that of C_3 plants.

22. Certain electrolytes are present at inhibitory concentrations for plant growth though photosynthesis is not necessarily their primary site of action. Salinity decreases productivity through reduced photosynthetic rates and leaf area production i.e. photosynthetic surface. At a conservative estimate, 400 km^2 of formally productive land is lost annually by secondary salinization.

23. To varying extents in different plant species, growth is limited in saline habitats at several levels of plant organization. Tolerance results from an ability to exclude salts from the sites of active metabolism. The effects of salinity on photosynthesis may be similar to those observed at low water potential, may be caused by induced nutrient deficiency, particularly of K^+/Mg^+ which are vital for stomatal opening and thylakoid-stroma ion gradients respectively, or may be caused by direct toxicity.

24. Nitrogen is important to photosynthetic production at several levels in the plant. It is taken up as ammonium ions (NH_4+) or nitrate (NO_3-) and by direct fixation of atmospheric nitrogen by the nitrogenase enzyme of symbiotic micro-organisms, and a significant proportion of its further metabolism may occur as photosynthetic reactions within the chloroplast. In general, C_4 plants show a higher efficiency of nitrogen use in dry-matter production. Deficiencies of nearly all the essential nutrients reduce the photosynthetic rate of higher plants.

25. Symbiotic nitrogen fixation still remains the major route for the incorporation of nitrogen into organic matter. A high metabolic energy requirement from respiration for the reduction of elemental nitrogen is required. Nitrogen reduction may therefore be limited by the supply of photosynthate under field conditions.

26. Some pollutants occur naturally at low levels in the environment but over the last few decades their concentration has increased, around and even away from industrial sites and cities, to levels which have deleterious effects on biomass production.

27. CO_2 is today a significant atmospheric pollutant. The present annual increase in atmospheric concentraiton is 0.7 mg kg (air)$^{-1}$ yr $^{-1}$. Current knowledge suggests that photosynthesis and biomass production will rise as a result of this increase, more in C_3 than C_4 plants, in situations where other factors e.g. water and nutrients are not primarily limiting plant growth.

28. Sulphur dioxide, nitrogen oxides and ozone enter plants largely through stomata. Changes in stomatal resistance are often observed but their primary effects are probably on photosynthetic processes in the chloroplast as a result of damage to cell membranes. A major effect of nitrogen oxide however may be to saturate nitrite reductase and thereby cause a build-up of nitrite to toxic levels.

29. Current evidence suggests that heavy metals and ultra-violet (UV) radiation inhibit several parts of the photosynthetic process. Disruption of the

thylakoid membrane and stromal lamellae may be responsible for the effects of UV. Pollutants may be present as mixed contaminants and have additive or synergistic effects on photosynthesis. Species differ in their sensitivity to exposure to pollutants and in some instances C_4 plants may differ from C_3 plants.

30. The capture of photons and their utilization in electron transport form the largest single area of research in photosynthesis. The two reaction centres, PSI and PSII constitute less than 1% of total chlorophyll and excitons are funnelled towards them from antennae chlorophyll and a light harvesting chlorophyll-protein complex (LHCP). The phosphorylation state of LHCP may determine the availability of trapped light to each photosystem and balance the requirements for NADPH and ATP.

31. The synthesis of carotenoids, superoxide dismutases, peroxisomes and catalases may be crucial to the protection of chlorophyll during periods of stress and low CO_2 availability when molecular oxygen may act as an electron acceptor for PSI leading to the production of the extremely reactive free hydroxyl radicals.

32. Light is limiting to the productivity of an established crop. It is not clear whether this is due to insufficient production of NADPH and ATP or to indirect control of the dark reactions via the light reactions.

33. Some enzymes of carbon metabolism are induced by changes in the reducing state of the stroma in the presence of light. This causes a lag in the induction of CO_2 assimilation following dark-light transitions. Studies of isolated chloroplasts suggest that such lags might also result from the time required to raise the level of intermediates of the RPP-cycle to a maximum through the autocatalytic property of the cycle. The activity of the rate-limiting enzymes and the amount of substrate must influence photosynthetic rates and therefore biomass production.

34. CO_2 limits the rate of photosynthesis in single leaves and crops under many conditions. The dark reactions of photosynthesis therefore will limit the production of dry matter independent of the supply of NADPH and ATP from the light reactions.

35. Photorespiratory losses of CO_2 can amount to 20-60% of total biomass production in C_3 plants. At present, decreasing photorespiration by increasing the atmospheric CO_2 concentration around crops or by altering the affinities of Ru*bis*CO for its substrates CO_2 and O_2 are not practicable ways of increasing biomass production on any large scale.

36. Gas-exchange analysis has been the most important technique for measuring the photosynthetic performance of single leaves or crops. Controlled environments have been used to identify the mechanisms involved in the response of single leaves to light, temperature, carbon dioxide and oxygen, water and nutrient status, and other plant and environmental factors. Problems arising from inadequate simulation of field conditions in controlled environments make it difficult to establish how realistic these findings are of actual field behaviour.

37. Community gas-exchange of CO_2 has been measured by micrometeorological techniques, by the enclosure of groups of plants, by cuvettes and by labelled carbon dioxide. As many environmental factors are partially or wholly related, the interpretation of such measurements in terms of discrete effects of single

environmental variables can be problematical.

38. The interception of radiation by crop canopies and the length of the growing season are the major determinants of the maximum quantity of light which can be intercepted. The relationship between canopy photosynthesis or biomass production and light interception is linear until canopy closure at which time mutual shading of leaves ensures a continued increase in biomass production with increase in the amount of light.

39. Canopy structure, and in particular leaf orientation, are potentially important factors determining biomass production. The advantages of an erect leaf habit should be more pronounced at high leaf area indices and in C_3 cereals and forage grasses but to date these have only contributed in the successful further development of one crop, rice. Other canopy characteristics may be more important determinants of canopy structure and in practice differences in leaf angle may be less significant than differences in the rate at which the canopy expands to form a complete cover.

40. Growth analysis has provided quantitative measurements of actual biomass production over chosen intervals of time. The production of leaf area in terms of leaf area index and the persistence of this green area, is a more important determinant of biomass production than the unit leaf rate.

41. The product of photosynthesis and leaf area determines the total production of dry matter and not the individual leaf photosynthetic rate per se. Economic yield is also determined by the partitioning and use of photosynthate and by controls which limit photosynthesis through the rate of translocation.

42. Models have been developed to simulate the photosynthetic performance of biomass production of plants over a wide range of conditions. Mechanistic models have provided a link between sub-cellular processes, gas-exchange and production research, and sophisticated light distribution models have been used as the basis for modelling canopy photosynthesis.

Yield prediction has been the primary objective of other models and can be used to demonstrate the contribution of photosynthesis as a major determinant of productivity.

43. Increased food and biomass production are essential in a world where 600 million people are estimated today to be seriously undernourished and hungry and where biomass is also the main source of fuel, clothing fibres and building materials.

44. The deleterious effects of man's activities on the environment through pollution, desertification and deforestation have raised serious questions on the world's ability to sustain increased photosynthetic production.

45. The major improvements in crop yield to date have come from improved fertilization of the land, improvements in pest resistance, pest protection and harvest index. Total dry-matter production by a crop imposes the limit on economic yield improvement by these techniques. As yet little attention has been paid to the possibility of increasing total dry-matter production through photosynthesis.

46. Increased leaf canopy photosynthesis is the key to increased plant production and a good understanding of both leaf development and photosynthetic

production processes is essential for further substantial yield improvements since it is leaf canopy photosynthesis which ultimately sets the upper limit on improved bioproductivity.

PART I
Photosynthesis and Biomass

Introduction

The improvement of crop productivity to date has resulted largely not from an overall increase in crop photosynthesis, but an increase in the amount of the total photosynthate invested into the portion of the plant of economic value, e.g. the grain of the wheat plant, the fibres of the cotton seed, and the trunk of the pine tree. This has been achieved by improved harvest index and improved disease and pest resistance. In instances where crop photosynthesis has been enhanced by breeding, it is usually as a result of improvement of the plant's ability to produce a leaf canopy, i.e. genotypes capable of producing increased photosynthetic area in response to inorganic fertilisers, and not through the efficiency of photosynthesis itself at an individual leaf level. However, photosynthetic capacity sets the limit on improvement of yield and ultimately man must be able to improve photosynthetic capacity if further increases in crop yield are going to be achieved when other limiting factors are eliminated.

Photosynthesis has been one of the most widely researched topics in the area of plant sciences and the outcome is quite a good understanding of the photosynthetic process at the cellular level. Photosynthesis has also been measured extensively at the whole plant level in the laboratory and in the field. There are only a few instances however where any attempt has been made to integrate photosynthetic studies with biomass production. This book attempts to bring together the existing information for terrestrial environments and considers the current role of photosynthesis in biomass production.

CHAPTER 1
Photosynthetic Mechanisms

1.1 PHOTOSYNTHESIS AS A PROCESS

As implied in its name, photosynthesis means literally the assembly of a product from raw materials using light. In this instance the products are carbohydrate (ultimately biomass) and oxygen, the raw materials carbon dioxide and water, and the means, sunlight and green plants. The chemical reaction can be simply represented as a redox (reduction-oxidation) reaction, in which carbon dioxide is reduced to carbohydrate and water oxidized or "split" to form oxygen, by the equation,

$$CO_2 + 2H_2O \xrightarrow[\text{green plants}]{\text{sunlight}} [CH_2O] + O_2 + H_2O$$

In addition to C, H and O, plants also incorporate N and S into their organic structure via photosynthetic reactions. Although, the net result of the photosynthetic reduction of CO_2 may be summarized in a simple equation, the whole photosynthetic process is highly complex and made up of a sequence of partial processes which span time scales from 10^{-13} s (light harvesting processes) to 10^7 s (bioproductivity). This chain of processes also incorporates many feedback control mechanisms. There is as yet no well-defined link between the extremes of the process but a better realization of the factors linking the initial light input to the final biomass output may be possible if the key areas, light reactions, dark reactions and the diffusion of carbon dioxide are first considered at their present level of understanding. Further, some aspects of the photosynthetic process are now known to differ fundamentally between groups of green plants and have been shown to relate importantly to differences in environmental tolerance and productivity.

1.2 LIGHT REACTIONS

The "light reactions" of photosynthesis encompass light harvesting, i.e. the primary photochemical act, electron transport and photophosphorylation. They occur in or on the thylakoid membranes of the chloroplast and in this respect are spatially separated from the enzymes of photosynthetic carbon (and nitrogen) metabolism which are localized in the stroma. The fundamental aspects of the light reactions are summarized in Fig. 1.1.

The light reactions of photosynthesis are located in the thylakoids and stromal lamellae. The thylakoids are flattened disc-like vesicles which usually stack to form grana and are interconnected by the stromal lamellae. A single thylakoid disc of about 0.5 μm diameter may contain 10^5 chlorophyll and associated pigment molecules. These pigments include chlorophyll *a* which is ubiquitous in all photosynthetic organisms capable of splitting water in photosynthesis. Chlorophyll *a* assumes a number of forms within the thylakoid membrane. These are expressed as changes in absorption/fluorescence spectra that presumably depend on the protein molecules with which the chlorophyll molecules are associated and their "solvent" environment (Seely, 1977; Thornber and Alberte, 1977). Several other pigments, chlorophyll *b*, carotenes and xanthophylls are also present and are capable of passing energy from absorbed photons to chlorophyll *a*.

Fig. 1.1 In higher plants and all other photosynthetic eukaryotes, the chloroplast is the site of the conversion of intercepted light energy to stored chemical energy, principally via the reduction of CO_2. (a) A 3-dimensional view of chloroplast structure showing the major structural features of a fully developed higher plant chloroplast. (Reproduced with permission from Reid and Leech, 1980.) Functionally the chloroplast may be divided into two parts the chloroplast internal membranes and the stroma. (b) A hypothesized arrangement of the photosystems and intermediates of electron transport within the chloroplast membranes. Note that PS2 is limited to the appressed areas of the membranes, i.e. between adjacent surfaces of the granal stacks, whilst PS1 and the coupling factor (CF_0–CF_1) occur in the non-appressed regions, i.e. the exposed edges of the granal stacks and the stromal thylakoids. In the membrane, trapped light energy drives the flow of electrons from water through the electron transport intermediates terminating in the reduction of NADP. The passage of electrons both to NADP and in cyclic electron transport translocates protons from the stroma to the space enclosed by the membranes. This gradient provides the electrochemical potential energy for ATP synthesis at the CF_0–CF_1. Arrows indicate the direction of flow of electrons and protons. Abbreviations are: Photosystem 2 (PS2), photosystem 1 (PS1), light harvesting chlorophyll-protein complex (LHCP), reaction centre molecule of PS2 (P680), primary acceptor of P680 (Q), plastoquinone/plastoquinol (PQH2/PQ), plastocyanin (PC), reaction centre molecule of PS1 (P_{700}), primary acceptor of P_{700} (X), Rieske iron-sulphur protein (FeS), cytochromes b_6 and f (Cyt b_6, Cyt f) and Ferredoxin (Fd) (reproduced with permission from Barber, 1983). (c) CO_2 reduction occurs in the stroma via the RPP cycle which is driven by the ATP and reduced NADP generated through the membrane electron transport. The number of lines per arrow indicate the number of times each reaction must occur for one complete turn of the cycle in which three molecules of CO_2 are assimilated and reduced. The dashed lines indicate the principal reactions removing intermediates from the cycle for biosynthesis. Abbreviations are: ribulose bisphosphate (RubP), Glycerate 3-phosphate (PGA), Glycerate bisphosphate (bPGA), Glyceraldehyde 3-phosphate (GA3P), Dihydroxyacetone 3-phosphate (DHAP), Fructose bisphosphate (FbP), Fructose 6-phosphate (F6P), Sedoheptulose bisphosphate (SbP), Sedoheptulose 7-phosphate (S7P), Erythulose 4-phosphate (E4P), Xylulose 5-phosphate (Xu5P), Ribose 5-phosphate (R5P), Ribulose 5-phosphate (Ru5P) and Thiamine pyrophosphate (TPP) (reproduced with permission from Bassham and Buchanan, 1983).

PHOTOSYNTHETIC MECHANISMS

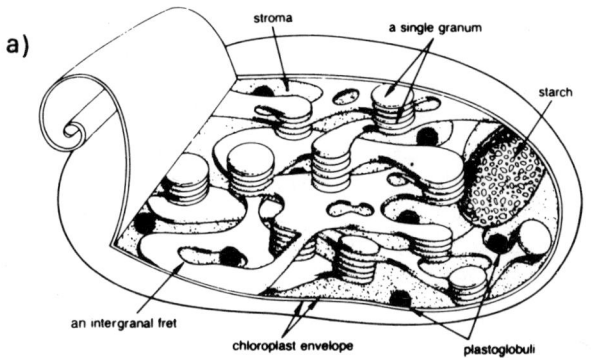

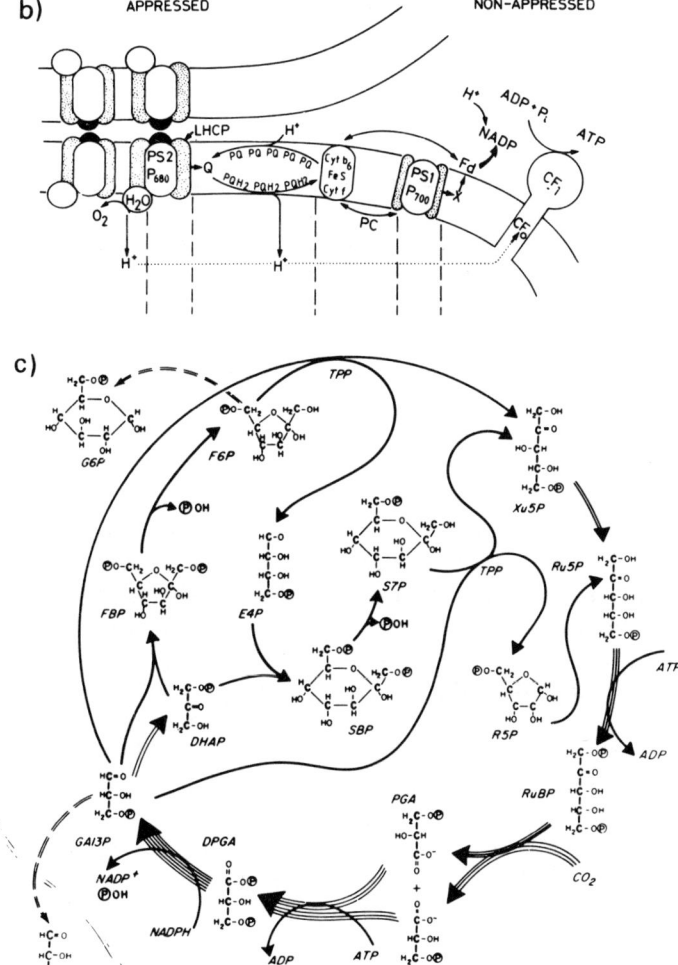

The pigment-protein complexes are organised into two photosystems, I and II (PSI and PSII), and a light harvesting complex (LHCP) which can transfer excitons (the discrete units of energy gained from absorption of photons) to the photosystems and from PS II to PS I by "spill-over". The amount of such "spill-over" has been shown to change with time during induction of photosynthesis and is regulated both by ion gradients across the membrane and by ATP concentrations (Barber, 1977; Bennett *et al.*, 1980). This allows a proper distribution of captured photons, an essential feature for obtaining maximum efficiency of light energy conversion (Barber, 1982). Once captured within the membrane, the energy of a photon can be lost by radiationless decay or by fluorescence, or it can be used to drive a chemical process. Within the photosystems, excitons are transferred through a "random walk" between chlorophyll molecules until they reach the reaction centre molecule, P700 in PS I and P680 in PS II.

The second stage of the "light reactions" *viz* electron transport occurs also within the photosynthetic membranes and is quite well understood with some notable exceptions. In PS II a manganese-containing complex is oxidised by P680 each time it receives an exciton, and in the process of losing four electrons it releases oxygen from water (Fig. 1.1). Concurrently, electrons are transferred ultimately from the ater, to the primary acceptor of P680, "Q". Neither the exact mechanism of water-splitting nor the identity of Q are known. The unknown Q transfers its electron through "B" to plastoquinone (PQ) which binds with two protons on the thylakoid outer surface. Reduced plastoquinone (PQH_2) is then oxidised on the thylakoid inner surface releasing protons to the inner space (Fig. 1.2). This mechanism combined with the water splitting helps to create a proton gradient across the membrane for which Mg^{2+} apparently provides a counterion. Electrons from PQ are then transferred via cytochrome f through plastocyanin to P700 which utilizes the energy of a second exciton from PS1 to pass an electron via membrane bound Fe-S centres to ferredoxin, NADP reductase and NADP (Fig. 1.1). The reduced NADP (NADPH) forms the reducing power for the next stage of photosynthesis, i.e. CO_2 assimilation via the "dark reactions".

The energy consumed in the generation of a proton gradient created by electron transport is utilized for the synthesis of ATP by ATPase/ATP synthase (CF_1) particles on the exposed surfaces of the grana and stromal lamellae (Golbeck *et al.*, 1977). Protons diffuse back into the stroma via specific channels (CF_0) in the membrane which are associated with the synthase particles where the passage of protons results in the synthesis of ATP, probably in the ratio of $3H^+:1ATP$ (Fig. 1.1; Shavit, 1980). Besides this pathway of non-cyclic electron transport, cyclic electron transport also allows the production of a proton gradient and thus photophosphorylation. Cyclic electron transport utilizes only PS I but is primed with electrons from PS II. The same photosystem 1 units are probably used for both cylic and non-cyclic electron flow. Stromal lamellae are enriched in PS I. During cyclic electron flow, excitons passed to P700 causing electron flow to ferredoxin NADP reductase are passed via unknown carriers to rejoin the pathway used for non-cyclic electron flow to P700 (Figure 1.1). The result is proton pumping into the thylakoid inner space by a shuttle involving a quinone-protein complex which is as yet only partially characterized. Thus, ATP can be generated in a cyclic manner from a proton gradient established without water splitting. ATP from either non-cyclic or cyclic electron transport forms the energy source for the "dark-reactions" but is also available for other metabolic processes within the chloroplast, such as protein synthesis.

1.3 DARK REACTIONS

The light energy trapped as NADPH and ATP by chloroplasts can be utilized in many processes. The stoichiometry of CO_2 assimilation and O_2 evolution by mature leaves suggests that CO_2 assimilation is the main "sink", but nitrogen and sulphur metabolism clearly also use significant amounts of this reducing power and phosphorylating energy (Lea and Miflin, 1979; Schmidt, 1979).

The Calvin or reductive pentose phosphate cycle (RPP-cycle) appears to be ubiquitous to plants and is the only form of CO_2 carboxylation which results in a net production of dry matter. The cycle follows a reaction pathway first elucidated by Calvin and co-workers 30 years ago (Bassham, 1979). The principal step which transfers energy from the products of photochemistry to carbohydrates is the phosphorylation and reduction of glycerate 3-P (PGA) to glyceraldehyde 3-P (Figure 1.1). The enzyme responsible for the carboxylation of CO_2 with its substrate ribulose 1,5-bisphosphate (RubP) is RubP carboxylase (RubisCO) and the product two molecules of PGA. The 3-carbon glyceraldehyde is either used to regenerate the 5-C substrate RubP through a complex series of reactions which requires a further input of ATP or leaves the cycle as a 6-carbon sugar (Fig. 1.1). This sugar, of course, is the source of dry matter production. From the constituent parts of the RPP-cycle, it can be seen that 3 ATP and 2 NADPH are required for each complete turn of the cycle. This compares to the ATP:NADPH production ratio during non-cyclic electron transport of 1.0 to 1.5. The whole topic of photosynthetic carbon metabolism and related processes has been extensively reviewed (Gibbs and Latzko, 1979; Hatch and Boardman, 1981; Edwards and Walker, 1983).

1.4 PHOTORESPIRATION

A second metabolic pathway influencing photosynthesis is the C_2 or glycollate pathway which ultimately results in the loss of previously fixed CO_2 from the plant.

Ru*b*P carboxylation is the ubiquitous process by which all higher plants assimilate CO_2 in photosynthesis but a universal property of Ru*bis*CO appears to be its ability to catalyse both oxygenation and carboxylation of Ru*b*P (Lorimer, 1981). Oxygenation of Ru*b*P produces one molecule of PGA which is metabolized through the RPP cycle and one molecule of glycolate-2P which is metabolized through the C_2 pathway. In the photosynthetic C_2 pathway two molecules of glycollate are metabolized via a series of reactions in the peroxisomes and mitochondria to one molecule of CO_2 and one molecule of PGA (Figure 1.2; Tolbert, 1979). The net result of Ru*b*P oxygenation and glycolate-2P metabolism is the light dependent evolution of CO_2 and consumption of O_2. The manifestation of this process in gaseous fluxes is termed photorespiration.

Photorespiration may be an essential regulatory mechanism for the removal of excess photoreductant from the chloroplast under conditions where CO_2 is not available as the terminal electron acceptor. Alternatively it may have arisen by accident because Ru*bis*CO catalyzes an oxygenation reaction which results in the loss of glycollate. Photorespiration could then be seen as a metabolic cycle which retrieves part of the carbon lost from the Calvin cycle in the reduced C_2 form.

1.5 THE DIFFUSION PROCESS

In spite of its importance as a major substrate, CO_2 is present in the atmosphere in quite small concentrations and is limiting to photosynthesis in full sunlight. CO_2 enters the leaf because a diffusion gradient exists between the sites of photosynthesis and the atmosphere. Thus, the net rate of photosynthetic CO_2 assimilation is directly related to the rate of CO_2 flux through this diffusion gradient. The flux, F, is determined by the size of the gradient. As the flux of gases between regions of different concentration is analogous to the flow of electricity through an electrical conductor, by analogy to Ohm's law:

$$F = \Delta C / \Sigma r \quad \ldots\ldots\ldots\ldots(1.1)$$

where the flux of CO_2 into the leaf (F), the concentration gradient (ΔC) and total resistance of the leaf to CO_2 diffusion (Σr) are analogous to the current, potential difference and electrical resistance, respectively.

The idea of describing the process of CO_2 assimilation through a resistance analogue was developed by Gaastra (1959) who considered that the pathway of CO_2 diffusion between the atmosphere and point of carboxylation consisted of three resistances in series: the boundary-layer resistance (r_a), the stomatal resistance (r_s) and the mesophyll resistance (r_m) (Fig. 1.3). It followed that:

$$F = (C_a - \Gamma)/(r_a + r_s + r_m) \ldots\ldots\ldots\ldots(1.2)$$

PHOTOSYNTHETIC MECHANISMS

Fig. 1.2 The photosynthetic C_2 or glycollate pathway showing the catalysis of RubP oxygenation by RubisCO to form one molecule of phosphoglycollate and one molecule of PGA. Carbon dioxide is released during the conversion of two molecules of glycine to one of serine in the mitochondria. Numbers indicate the enzymes involved in each step of the pathway: 1, ribulose-P_2 carboxylase/oxygenase (RubisCO); 2, P-glycolate phosphatase; 3. NADH glyoxylate reductase; 4. glyoxylate oxidation by any oxidant; 5, glycolate oxidase; 6, catalase; 7, glutamate-glyoxylate aminotransferase; 8, serine-glyoxylate aminotransferase; 9, glycine oxidase; 10, serine hydroxymethyl transferase; 11, NADH-glutamate dehydrogenase; 12, glutamate-hydroxypyruvate aminotransferase; 13, glutamate-oxaloacetate aminotransferase or asptartate aminotransferase; 14, NADPH-glutamate dehydrogenase; 15, NADH-hydroxypyruvate reductase or glycerate dehydrogenase; 16, NAD-malate dehydrogenase; 17, NADP-malate dehydrogenase; 18, glycerate kinase; 19, P-glycerate phosphatase (reproduced with permission from Tolbert, 1979).

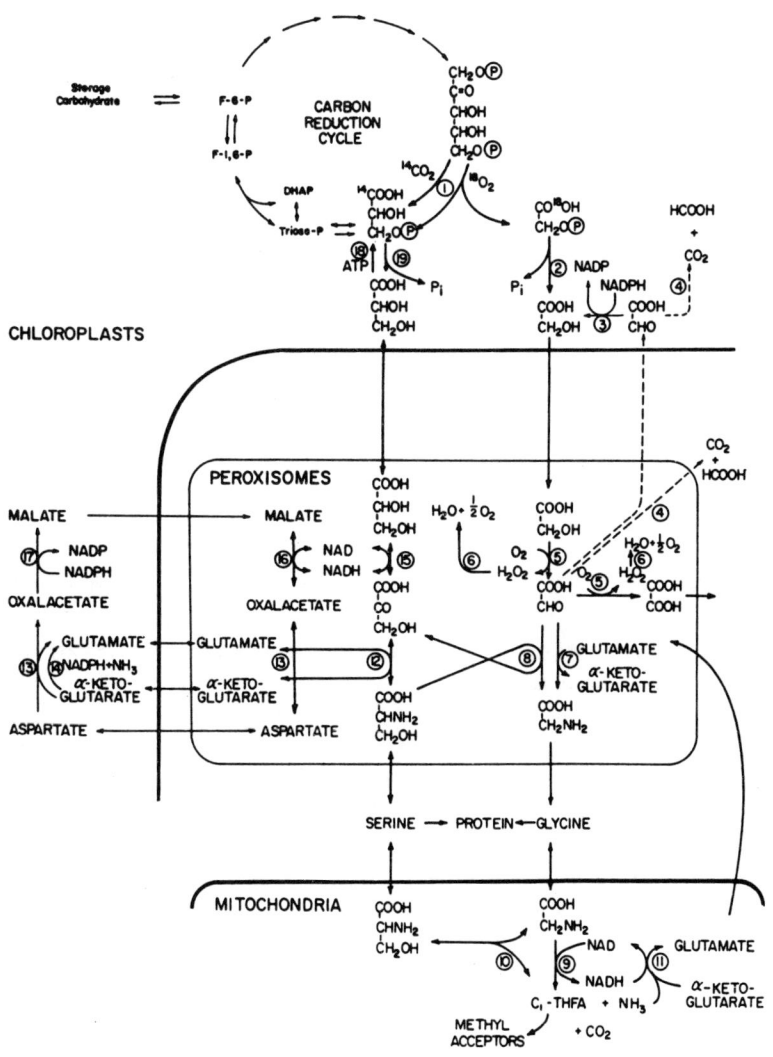

where C_a is the CO_2 concentration in the atmosphere and the CO_2 concentration at the site of carboxylation unknown, but assumed to approach zero in Gaastra's model, i.e. $\Delta C = C_a$. In later models the CO_2 compensation point of photosynthesis (Γ) has been considered a better estimate of the concentration inside the leaf.

1.5.1 Boundary layer resistance

When a gas passes over a flat surface such as a leaf there is a small non-turbulent layer of air molecules associated with the surface. This is the boundary layer. The depth depends on the geometry of the surface and on the velocity of the gas flowing over the surface. If the layer is deep, e.g. over a large leaf surface or in still air, the resistance to gas diffusion is greater, the diffusion of water for CO_2 into and out of the leaf is slower, and the resistance (r_a) is larger. Therefore, r_a is decreased by increase in wind speed and decrease in leaf size. The magnitude of r_a is generally a small fraction of the total resistance.

1.5.2 Stomatal resistance

The diffusive resistance encountered by CO_2 entering the leaf is largely proportional to the stomatal aperture and represents the variable resistance in the non-metabolic part of the photosynthetic CO_2 assimilation pathway. Its value is a function of the light flux density, leaf temperature, CO_2 concentration, leaf water potential and leaf-air humidity deficit.

1.5.3. Mesophyll resistance

This resistance is really a combination of resistances which include physical diffusion resistances to movement of CO_2 in the air spaces and from the mesophyll cell walls to the sites of carboxylation within the cell together with biochemical rate limitations. Its value therefore depends on the efficiency of cellular transfer of CO_2 and the efficiency of cellular transfer of CO_2 and the efficiency of the light and dark reactions in the chloroplasts. Limitations to photosynthesis by CO_2 in the mesophyll can be analysed further with respect to the biochemical limitations in terms of a "carboxylation efficiency", a measure proportional to the amount of *Rubis*CO in the leaves of C_3 plants (Farquhar and Sharkey, 1982).

Experimental procedures for the determination of these resistances are described by Long (1982) and given extensive analysis by Jarvis (1971). Resistance models allow numerical evaluation of limitations to CO_2 diffusion and thus productivity, given that supply of CO_2 is limiting photosynthesis. Thus, if $r_s = 250$ s m^{-1} and $\Sigma r = 500$ s m^{-1} it can be deduced that r_s accounts for 250/500 or one-half of the sum of limitations to CO_2 assimilation in that leaf under the conditions of measurement.

Fig. 1.3 Diagram illustrating a resistance analogue model of CO_2 diffusion into the leaf. C = CO_2 concentration (C_a = in ambient air, C_s = at the stomatal pore entrance, C_w = at the liquid/air interface of the mesophyll, C_i, the intercellular CO_2 concentration, C_c = at the site of carboxylation). r = resistance to CO_2 diffusion (r_a = in the boundary layer, r_s = through the stomata, r_m = in the mesophyll). The flux of CO_2, F through the chain of resistors is $F = (C_a - C_s)/r_a = (C_s - C_w)/r_s = (C_w - C_c)/r_m = (C_a - C_c)/(r_a + r_s + r_m)$. C_c is usually considered to be roughly half of the CO_2 compensation point, Γ (reproduced with permission from Long, 1982).

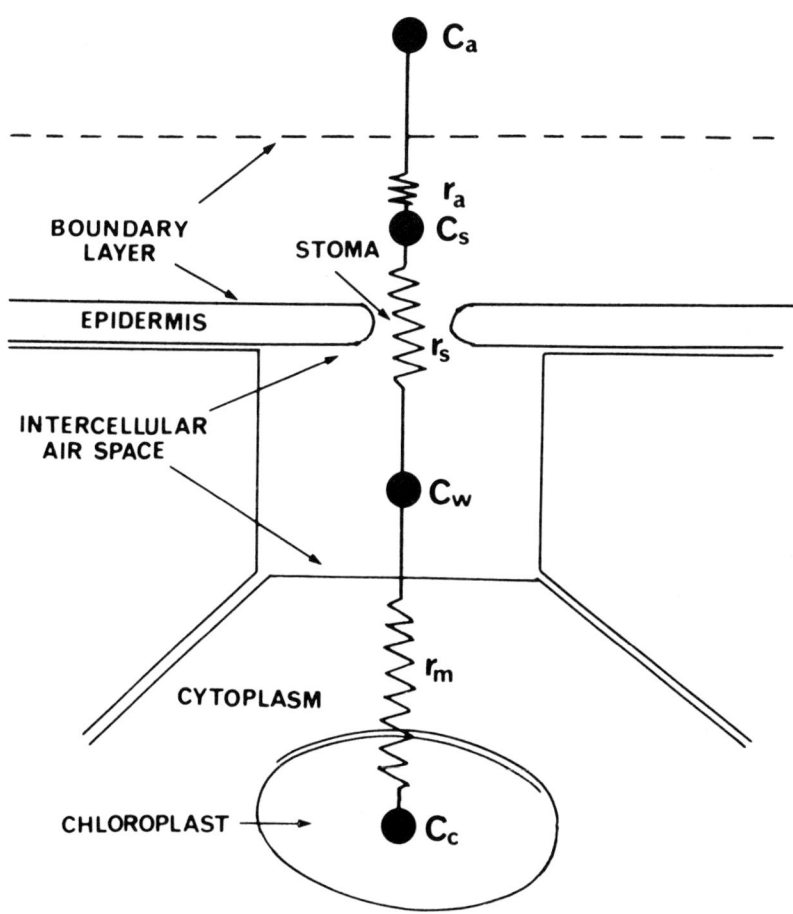

Typical resistances to CO_2 diffusion for a leaf of a mesophyte under optimal conditions would be in the ranges:

$$r_a = 10\text{--}30 \text{ s m}^{-1}$$
$$r_s = 250\text{--}1000 \text{ s m}^{-1}$$
$$r_m = 250\text{--}4000 \text{ s m}^{-1}$$

CHAPTER 2
Photosynthetic Diversity

2.1 C₃ PLANTS

While their photochemistry has remained a conservative characteristic of plants, their carbon metabolism has evolved more diverse forms and this feature has been used to distinguish photosynthetic as opposed to taxonomic groupings of species. Plants which reduce carbon dioxide solely through the RPP-cycle into the 3-C compound PGA are termed C_3 plants. The carbon dioxide is assimilated in the chloroplasts of the leaf mesophyll, but all of these chloroplasts are also capable of supporting photorespiratory processes (Tolbert, 1979). C_3 plants are the most widely distributed photosynthetic type and this photosynthetic mechanism is well adapted to all latitudes.

Two other types of carbon fixation incorporate the RPP-cycle and species using these pathways are identified accordingly as C_4 or CAM (Crassulacean acid metabolism) plants. A fourth group which has photosynthetic characteristics "intermediate" between C_4 and C_3 plants has also been suggested.

2.2 C₄ PLANTS

C_4 species initially fix carbon dioxide with phosphoenolpyruvate (PEP) into 4-C anions (malate or aspartate). The reaction is catalysed by PEP carboxylase (PEPc, Kortschak et al., 1965; Hatch and Slack, 1966). The initial fixation of CO_2 takes place in the mesophyll cells and the anions are then transported and decarboxylated in specialized bundle sheath cells which are photosynthetic. The released carbon dioxide is next refixed through the RPP-cycle which is exclusive to the bundle sheath cells (Hatch and Slack, 1970). The C_3 fragment resulting from decarboxylation of the organic acids diffuses back to the mesophyll cells where it is converted into phosphoenolpyruvate for the next carboxylation (Hatch, 1977).

The carbon metabolism of C_4 plants therefore differs from that of C_3 plants in several basic respects. Photosynthesis is spatially separated into two compartments.

One compartment, the mesophyll cells contains a non-autocatalytic process based on a distinctive enzyme complement. The second compartment, the bundle sheath cells, houses the autocatalytic RPP-cycle which is found in the mesophyll cells of C_3 plants and which is essential to C_4 plants for the net production of dry matter.

The characteristics which distinguish C_4 from C_3 plants are referred to as the C_4 "syndrome" (Tregunna et al., 1970). Three subtypes within C_4 species may be distinguished on the basis of leaf anatomy and biochemistry. These are the "NADP-me type" (NADP-malic enzyme species), the "PEP-ck type" (PEP-carboxykinase species) and the "NAD-me type" (NAD-malic enzyme species) (Hatch and Osmond, 1976; Hatch, 1982). This classification, based on the mechanism of decarboxylation used in the bundle sheath of each subtype, is summarized in Fig. 2.1.

Fig. 2.1 Outline of C_4 photosynthetic metabolism in the three subtypes of C_4 plants (NADP-me type, PEP-ck type and NAD-me type). The upper cells in the diagrams represent the mesophyll and the lower the bundle sheath. The malate-pyruvate shuttle from the mesophyll cells aplies to NADP-me type species and the aspartate-alanine shuttle to PEP-ck type and NAD-me type species. The enzymes involved are: (1) PEP carboxylase; (2) NADP malate dehydrogenase; (3) aspartate amino transferase; (4) alanine uminotransferase; (5) pyruvate, Pi dikinase; (6) adenylate kinase; (7) pyrophosphotase; (8) 3-PGA Kinase, NADP glyceraldehyde-3-P dehydrogenase and triose-P isomerase; (9) NADP malic enzyme; (10) PEP carboxykinase; (11) NAD malate dehydrogenase; (12) NAD malic enzyme (reproduced with permission from Hatch & Osmond, 1976).

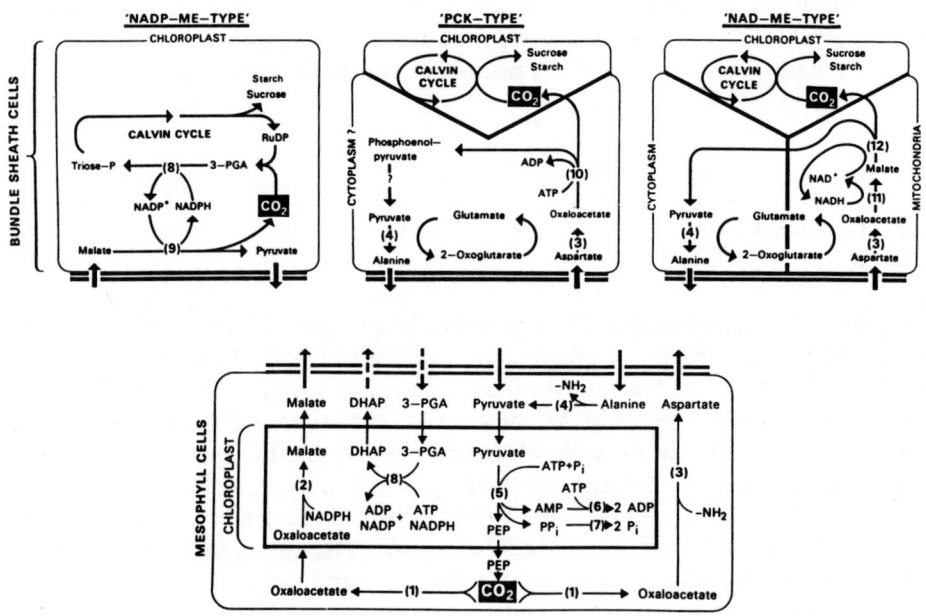

Anatomically, C₄ plants differ in leaf anatomy from C₃ species in possessing a thick walled sheath or rim (referred to as Kranz by Haberlandt, 1884) around the vascular bundles. The chloroplasts within the bundle sheath differ in size, shape and number from those of the surrounding mesophyll cells and their position as well as the position of the cells themselves (Hatch *et al.*, 1975; Hattersley and Watson 1975, 1976) can be used to characterize their decarboxylation reactions (Fig. 2.2).

Fig. 2.2 Subtypes of C₄ photosynthesis distinguished on the basis of leaf anatomy (a) in the family Graminae. Kranz cell (bundle sheath) chloroplasts in NADP-me type – centrifugal and agranal; PEP-ck type – centrifugal with grana; NAD-me type centripetal with grana (reproduced with permission from Edwards & Huber, 1981) (b) and in C₄ plants in general showing the relationship between their anatomy and biochemistry (reproduced with permission from Bolhár-Nordenkempf, 1982).

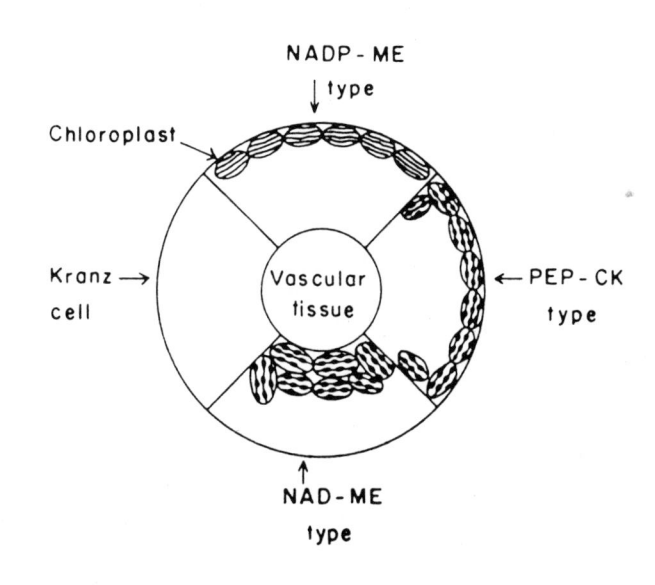

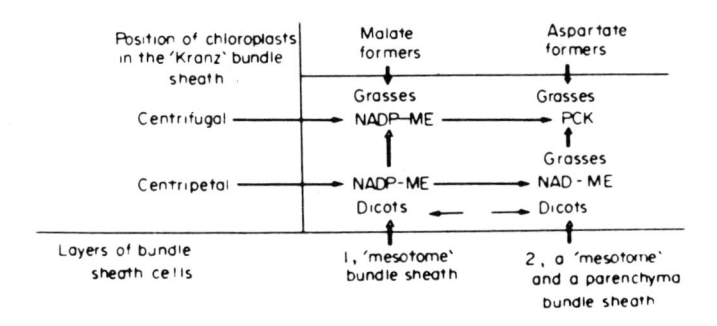

Compartmentation of photosynthesis in C_4 plants into a C_4 and C_3 step is essential to their function. The C_4 mechanism ensures that carbon dioxide is concentrated at the sites of fixation in the bundle sheath cells and potential losses of CO_2 from photorespiratory processes are avoided by its immediate refixation in the mesophyll cells (El-Sharkawy et al., 1968; Hatch and Kagawa, 1973). In essence, the outer layer of mesophyll cells which contains high activities of phosphoenolpyruvate carboxylase serve as the mechanism for concentrating CO_2 into the bundle sheath cells, thus decreasing O_2 competition for RubP and recycling any CO_2 released from the bundle sheath via glycolate metabolism. The end result is a higher net photosynthetic rate in C_4 than in C_3 plants. The C_4 syndrome is thought to have evolved, at least in part, as a response to the reduced efficiency of the RPP pathway of C_3 plants with increasing partial pressures of oxygen in the atmosphere (Smith, 1976). As an evolutionary process therefore, it has been described as an experimental approach by plants to cope with photorespiration and the syndrome appears to have evolved several times (Laetsch, 1974; Smith and Robbins, 1975). The possible causes of the appearance of C_4 plants in conditions which favour high photorespiration have been reviewed (Moore, 1981, 1983; Osmond et al., 1982).

High temperatures during the growing period favour growth of C_4 rather than C_3 species (Doliner and Jolliffe, 1979). Minimum temperature during the growth period was the climatic variable which correlated most closely with the distribution of C_4 species of the Graminae and Cyperaceae, though the occurrence of C_4 dicotyledons was more closely correlated with summer pan evaporation (Teeri and Stowe, 1976; Stowe and Teeri, 1978; Teeri et al., 1980). The occurrence of C_4 species may well be less defined than was at first thought, however, and in some instances well adapted to moist temperature conditions (McWilliam and Ferrar, 1974; Long and Woolhouse, 1978). Osmond et al. (1982) have pointed out that C_4 plants are not necessarily more tolerant of low moisture than are C_3 plants. In fact some desert environments, where extreme water stress is the major factor determining plant growth, are dominated by C_3 plants.

Available evidence also suggests distinctive distributional trends of the grass Kranz subtypes. A study of the geographical distribution of grasses in Namibia (Ellis et al., 1980) observed that malate formers (NADP-me) were more abundant with increasing rainfall whereas aspartate formers (NAD-me, PEP-ck) showed the opposite tendency. In general, however, C_4 species are ecologically more specialized than C_3 species and may possess competitive advantages under conditions of high temperature and irradiance and intermittent periods of drought which favour high rates of photorespiration (Doliner and Jolliffe, 1979). Woolhouse (1978) supports the general belief that the environmental forces favouring the emergence of the C_4 syndrome were certainly high temperature and light but that this did not preclude further adaptation of the syndrome into areas where selection was not strong enough to evoke it in the first instance.

2.3 CAM PLANTS

Plants which utilize Crassulacean acid metabolism fix carbon dioxide into the C_4 anion, malate and into glycerate 3-phosphate (PGA), but these carboxylations are separated in time rather than in space, as they are in the leaf of a C_4 species (Fig. 2.3; Kluge, 1979; Osmond and Holtum, 1981; Ting and Gibbs, 1982). The stomata of CAM species are open at night and malic acid accumulates in the vacuoles; during the day the stomata are closed, malate is decarboxylated and CO_2 enters the RPP-cycle. It is probable that CAM plants exhibit photorespiration, but it is difficult to detect.

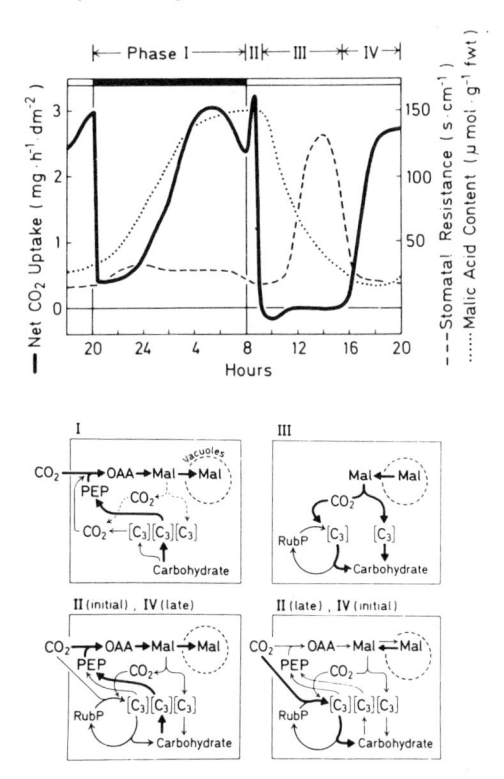

Fig. 2.3 Carbon fixation and flow in a typical CAM plant, Kalanchio claigremontia showing the separation of C_4 and C_3 carboxylations in time rather than space. Net CO_2 exchange (⎯⎯⎯); malate rhythm (...............) stomatal resiteance (----------). Phase I: carbon dioxide is fixed into malate during the dark period; Phase III: decarboxylation of malate and refixation of CO_2 into the RPP-cycle. Phases II and IV are transition periods which also involve the provision of carbon skeletons for conversion to PEP (Phase II) and a mixture of external CO_2 fixation by both pathways (C_4 and C_3) when endogenously supplied CO_2 no longer saturates the photosynthetic apparatus. $[C_3]$ = C-3 carbon skeletons; Mal = malate; OAA = oxaloacetate; PEP = phosphoenol pyruvate; RuBP = ribulose 1,5-bisphosphate (reproduced with permission from Kluge et al., 1982).

Because of their ability to close stomata during the day but at the same time fix CO_2 released from the nocturnally synthesized malic acid, CAM plants are well adapted to and typically occupy the arid areas of the world. CAM plants have extremely low transpiration ratios of (about 50-125 kg water transpired/kg CO_2 fixed) in contrast to C_4 and C_3 species (250–700), but are generally slow growing and their productivity is generally lower than that of C_3 or C_4 plants. Temperature and soil moisture affect the pattern of CO_2 assimilation since high temperatures and low soil moisture depress CO_2 assimilation (Lange et al., 1975). CAM plants have the capacity however to store water in their succulent organs and this permits the nocturnal opening of stomata.

The opening and closing of stomata is modulated by the CO_2 concentration in the mesophyll of CAM plants where stomatal closure during the day results from an increase in CO_2 concentration (up to 26%) during malic acid decarboxylation (Raschke, 1975; Ting and Gibbs, 1982). The conversion of well-watered CAM plants from C_3-type photosynthetic uptake during the day to 4-carbon acid production at night under severe stress conditions and vice-versa substantiates the hypothesis that the mechanism is an adaptation to severe water stress (Winter, 1974; Hartsock and Nobel, 1976; Winter et al., 1978). The physiological potential for storage of the organic acids and of stored carbohydrate for the synthesis of phosphoenolpyruvate result in the low photosynthetic rates, but the mechanism enables the plant to survive until more favourable conditions prevail (Woolhouse, 1978; Osmond et al., 1982).

Isotope discrimination ratios (δ ^{13}C) have provided a useful technique for the separation of C_4, C_3 and CAM species on the basis of their carboxylation reactions (Troughton, 1971; Smith, 1972; Vogel, 1980), though variation in δ ^{13}C values can occur within plants and in aquatic environments (Troughton, 1979). This ratio arises from the fractionation of carbon isotopes during carboxylation and is caused by preferential utilization of $^{12}CO_2$ and partial exclusion of $^{13}CO_2$ by the plant (Smith and Brown, 1973). A greater discrimination by Rubisco in C_3 plants results in lower ratios than for C_4 plants while the variability in ^{13}C ratios in CAM plants is evidence for a shift between C_4 and C_3 photosynthesis according to environmental conditions (Black, 1973; Osmond et al., 1973; Medina et al., 1977). The major distinguishing features of C_3, C_4 and CAM plants are summarized in Table 2.1.

2.4 "C_3/C_4 INTERMEDIATES"

In addition to the three groups of plants described above, a fourth group with C_3/C_4 intermediate anatomical and biochemical characteristics may be distinguished (Ogren and Chollet, 1982). In particular recent studies have shown that three Panicum species (P. milioides, P. decipiens and P. schenckii) exhibit these characteristics with respect to leaf anatomy and photorespiration (Kanai and

Kashiwaga, 1975; Kestler *et al.*, 1975; Brown and Brown, 1975; Morgan and Brown, 1979, Morgan *et al.*, 1980). It is suggested on the basis of leaf anatomy, rates of photorespiration, CO_2 compensation point and O_2 inhibition of photosynthesis that C_3/C_4 intermediate species occur in the genera *Mollugo* and *Moricandia* (Kennedy and Laetsch, 1974; Apel, 1980). The genus *Flaveria (Asteraceae)* has also been reported to possess species with C_3/C_4 intermediate characteristics. Other species which have been shown to possess the C_3/C_4 intermediate features include *Steinchisma hians, Alloteropsis semialata* and *Chamaesyce acuta* (Brown, 1977). Available evidence suggests that in C_3/C_4 species the leaf anatomy changes towards the Kranz architecture and this precedes biochemical changes towards C_4 photosynthesis. C_3/C_4 intermediates may provide a crucial evolutionary link between the C_3 and C_4 groups of plants.

2.5 OCCURRENCE

With very few exceptions, all sub-families, tribes and genera appear uniformly C_3 or C_4 plants within the Graminae (Smith and Brown, 1973) whereas C_4 dicotyledenous families encompass C_3 species which are also well adapted to arid conditions (Stowe and Teeri, 1978). The C_4 syndrome is known to be present in at least 18 angiosperm families (Acanthaceae, Aizoaceae, Amaranthaceae, Asclepiadaceae, Asteraceae, Boraginaceae, Capparaceae, Chenopodiaceae, Cyperaceae, Euphorbiaceae, Liliaceae, Molluginaceae, Nyctaginaceae, Poaceae, Polygalaceae, Portulacaceae, Scrophulariaceae and Zygophyllaceae). The angiosperm families possessing CAM species include: Agavaceae, Aizoaceae, Asclepiadaceae, Asteraceae, Bromeliaceae, Cactaceae, Crassulaceae, Cucurbitaceae, Didieraceae, Euphorbiaceae, Geraniaceae, Labiatae, Liliaceae, Oxalidaceae, Orchidaceae, Piperaceae, Portulacaceae and Vitaceae. Comprehensive lists of C_4 species are available (Downton, 1975; Raghavendra and Das, 1978; Imbamba and Papa, 1979; Smith, 1982), and include *Zea mays, Saccharum officinarum, Sorghum vulgare* and many tropical grasses. The major cereals, *Oryza sativa, Triticum aestivum* and *Hordeum sativum* are C_3 plants as are without exception all species from the *Leguminosae*. There are a few species of agricultural significance among CAM plants e.g. *Ananas comosus, Opuntia* spp., *Yucca* spp. and *Agave amaniensis*. Comprehensive lists of CAM species are provided by Szarek and Ting (1977), Szarek (1979) and Smith (1982).

Table 2.1 Some characteristics distinguishing C_3, C_4 and CAM plants (adapted from Black, 1973)

Characteristic	C_3	C_4	CAM
First stable product	C_3 compound (phosphoglycerate)	C_4 compound (aspartate and malate)	C_4 and C_3 compound (night and day respectively)
Leaf anatomy in cross section	Diffuse distribution of organelles in mesophyll or palisade cells with similar or lower organelle concentrations in bundle sheath cells if present	A definite layer of bundle sheath cells surrounding the vascular tissue which contains a high concentration of organelles; layer(s) of mesophyll cells surrounding the bundle sheath cells	Spongy appearance. Mesophyll cells have large vacuoles with the organelles evenly distributed in the thin cytoplasm. Generally lack a definite layer of palisade cells.
Interveinal distance (monocotyledons only)	large (more than five chlorenchyma cells)	small (around chlorenchyma cells)	large (more than five cells)
Chloroplasts	similar in all tissues	dimorphic	similar in all tissues
Leaf chlorophyll a to b ratio	$2.8 \pm .4$	$3.9 \pm .6$	2.5 to 3.0
Theoretical energy requirement for net CO_2 fixation (CO_2 : ATP : NADPH)	1 : 3 : 2	1 : 5 : 2	1 : 6.5 : 2
Major leaf carboxylation sequence in light	RubisCO	PEP carboxylase then RubisCO	PEP carboxylase at night then RubisCO in the day
Leaf isotopic ratio ($\delta^{13}C$)	-22 to -34	-11 to -19	-13 to -34

Response to net photosynthesis to increasing light intensity at temperature optimum	Saturation reached at about ¼ to ⅓ full sunlight	Either proportional to or only tending to saturate at full sunlight	Uncertain, but apparently saturation is well below full sunlight
Optimum day temperature for net CO₂ fixation	15 to 25°C	30 to 47°C	35°
Maximum rate of net photosynthesis (mg CO₂/m² of leaf surface/s)	0.4 to 1.1	1.1 to 2.9	< 0.4
Leaf photorespiration detection: (a) exchange measurements (b) glycolate oxidation	Present Present	Difficult to detect Present	Difficult to detect Present
Photosynthesis sensitive to changing O₂ concentration from about 1% to 21%	Yes	No	Yes
CO₂ compensation concentration (ppm CO₂)	35 to 70	0 to 10	0 to 5 in dark: 0 to 200 with daily rhythm
Transpiration ratio (gm H₂O/gm of dry wt)	450 to 950	250–350	50 to 55
Minimum stomatal resistances (s m⁻¹)	50–200	200–400	—
Minimum mesophyll resistances (s m⁻¹)	300–800	50–150	—
Maximum efficiency of light energy conversion	4.3%	5.8%	—

CHAPTER 3
Photosynthetic Productivity

3.1 WORLD PRODUCTIVITY

The total net primary production of the world can be expressed in terms of total photosynthetic carbon gain (photosynthesis) less the respiratory losses. Present estimates suggest that net annual primary production amounts to 8×10^{10} t carbon which is fixed into 2×10^{11} t of organic matter (Table 3.1) representing 10% of existing stored biomass. It should be noted that forests are clearly the most important resource in this respect. Similar quantities of carbon are stored as atmospheric CO_2 or CO_2 in ocean surface layers as are stored in biomass (Table 3.1). The energy content equivalent to net annual primary production (3×10^{21} J) however is about 10 times the world's annual energy use and 200 times our (human) food energy consumption even though the efficiency of the photosynthetic process is a mere 0.1% over the whole surface of the world.

Table 3.1 Carbon balance of the world[1]

	Carbon t
Net annual primary production	
(a) Total	8×10^{10}
	(2×10^{11} t organic matter)
(b) Cultivated land only	0.4×10^{10}
Stored biomass	
(a) Total (90% in trees)	8×10^{11}
(b) Cultivated land only	0.06×10^{11}
Atmospheric CO_2	7×10^{11}
CO_2 in ocean surface layers	6×10^{11}
Soil organic matter	$10-30 \times 10^{11}$
Ocean organic matter	17×10^{11}

[1] See Hall (1979) for original references.

Terrestrial net annual primary production is 4.8×10^{10} t carbon ($\simeq 1.2 \times 110^{11}$ t organic matter) and the annual input from cultivated land, 0.4×10^{10} t carbon or 8% of the total. As the terrestrial surface area of the earth is 13.1 Gha (= 1.31×10^8 km^2) and cultivated land accounts for 1.5 Gha (arable only) i.e. slightly more than 10%, the loss of land to cultivation has led to a loss of primary production (Buringh, 1980, Table 3.2).

Table 3.2 Assessment of net primary production of the world[1]

Land use	Area Gha	Net primary production t ha^{-1}	Gt
arable	1.5	6	9.0
grass	3.0	8	24.0
forest	4.0	16	72.0
urban	0.6	5	3.0
fresh water	0.4	12	4.8
other	3.5	2	7.0
total			119.8

[1] From Buringh (1980)

Net primary production from arable land is less than half that from forest overall, though estimates for modern agricultural systems suggest that if the removal of nutrients by cropping is offset by inputs from fertilizers, net primary production is similar (Buringh, 1980). It should be noted that many of these data are calculated estimates and not measured quantities and may be imprecise (Hall, 1979).

Annual food production is 1426 Mt (or 16% of total net primary production of arable land) of which 1126 Mt or 80% is accounted for by grains (Table 3.3).

Table 3.3 Annual food production

Source	Dry matter Mt	%
wheat	306	21
rice	272	19
maize	255	18
other grain	293	21
total grain	1126	79
tubers	106	7
sugar	12	1
other	68	5
total	1312	92
meat/milk	93	7
fish	21	1
grand total	1426	100

[1] From Buringh (1980)

There are considerable regional differences in production however (Buringh, 1980). The pattern of world grain trade is a reflection of this pattern (Table 3.4). Only two regions, North America and Australia/New Zealand have remained net exporters of grain throughout a 20-year period from 1960-1980, and apart from Western Europe there has been a considerable increase in the import requirements in other regions which will probably increase in future (Brown, 1981). In fact, only 12% of the world's cereal production enters world trade and a considerable proportion of the total primary food production (52%) including one-third of the world's grain production is converted to animal protein, particularly in developed countries. This reduces the total food energy available to man by approximately 40% (Hall, 1983).

Table 3.4 *The changing pattern of world grain trade*[1]

Region	Grain exports (+) and imports (−) (Mt)		
	1960	*1970*	*1980*[2]
North America	+ 39	+ 56	+ 131
Latin America	0	+ 4	− 10
Western Europe	− 25	− 30	− 16
Eastern Europe and U.S.S.R.	0	+ 1	− 46
Africa	− 2	− 5	− 15
Asia	− 17	− 37	− 63
Australia and New Zealand	+ 6	+ 12	+ 19
Total reserves	234	236	151

[1] From Brown (1975, 1981)
[2] World grain trade in 1980 was about 150 MT. In 1981 the USA exported one-third of its total production (110 Mt) or 55% of the world grain trade.

If world population continues to increase at a rate of 2% per annum and growth remains approximately exponential, it can be predicted that the population will double every 35 years (Allaby, 1977). Current predictions suggest that the world population will be 5276 millions in 1990 and is expected to eventually stabilize at a figure of more than 10 billion (Maudlin, 1980; Barr, 1981). In spite of this severe burden on demand, world grain production has increased faster to permit an annual improvement in per capita consumption approaching 1% per annum since 1945 (Sanderson, 1975). However, compound annual interest rates of increase in per capita food production have been decreasing over the last three decades since 1950 from 1.6% to 0.6% to 0.3% (Hall, 1983). The 3% annual improvement in production has been similar in developing and developed countries, but the much faster growth of population in the former, 2.5% per annum, has led to very little real progress towards self sufficiency in developing countries.

Buringh (1980) has concluded that the productive capacity of the biosphere is limited by cultural, socioeconomic and political conditions. Present estimates suggest that at least twice as much land is available for food production than that currently used (Wittwer, 1975). Should these constraints, including limitations by mineral deficiencies, plant diseases and farm management practices be removed it is speculated that maximum photosynthetic production could reach 40,000 Mt or 30 times that of present food production (Buringh et al., 1975).

3.2 NATURAL ECOSYSTEMS

3.2.1 Definition

Since all ecosystems have probably in some way been influenced by man, none can strictly be considered natural. In the context of this review the term natural ecosystem will be limited to land which has not been cultivated or planted by man.

Two measures of production must be distinguished. Gross primary production (P_G) is the photosynthetic assimilation of organic matter by a plant community during a specified period, most commonly one year, including the amount used by respiration. Net primary production (P_N) is gross primary production less respiratory losses (R_T, Eqn. 3.1). In an ecological context P_G is the sum of the photosynthetic inputs, R_T is the sum of the respiratory losses and P_N is the total photosynthetic input available to other trophic levels. The abbreviated term "production" in this section will refer to P_N.

$$P_N = P_G - R_T \quad \ldots\ldots\ldots\ldots (3.1)$$

3.2.2 Areas and Productivity

The areas occupied by the different ecosystems of the world and their mean, maximum and total productivities are given in Table 3.5. There have been many reviews of the net primary productivity of the biomes of the world (See Cooper, 1975; Rodin and Basilevic, 1966; Lieth and Whittaker, 1975; Holdgate et al., 1982). The two groups of ecosystems of immediate potential and actual importance to man through their primary production are the natural forests which occupy 4.0×10^7 km² with an annual rate of dry matter production amounting to 72 Gt yr⁻¹ and the natural grasslands which occupy 3.0×10^7 km² with an annual dry matter production rate of 24 Gt yr⁻¹ (Earl, 1975; Buringh, 1980; Table 3.2).

3.2.3 General

The reviews cited above and elsewhere which purport to show the net primary productivity of the biosphere are based on very limited data: they are essentially

Table 3.5 Net primary production and related characteristics of biosphere[1]

		Net primary production (dry matter)			Biomass (dry matter)		
Ecosystem type	Area ($10^6 km^2$)	Normal range ($g\ m^{-2}\ yr^{-1}$)	Mean	Total ($Gt\ yr^{-1}$)	Normal range ($kg\ m^{-2}$)	Mean	Total (Gt)
Tropical rain forest	17.0	1,000–3,500	2,200	37.4	6–80	45	765
Tropical seasonal forest	7.5	1,000–2,500	1,600	12.0	6–60	35	260
Temperate forest:							
evergreen	5.0	600–2,500	1,300	6.5	6–200	35	175
deciduous	7.0	600–2,500	1,200	8.4	6–60	30	210
Boreal forest	12.0	400–2,000	800	9.6	6–40	20	240
Woodland and scrubland	8.5	250–1,200	700	6.0	2–20	6	50
Savanna	15.0	200–2,000	900	13.5	0.2–5	4	60
Temperate grassland	9.0	200–1,500	600	3.4	0.25	1.6	14
Tundra and alpine	8.0	10–400	140	1.1	0.1–3	0.6	5
Desert and semi-desert scrub	18.0	10–250	90	1.6	0.1–4	0.7	13
Extreme desert – rock sand, ice	24.0	0–10	3	0.07	0–0.2	0.02	0.5
Cultivated land	14.0	100–4,000	650	9.1	0.4–12	1	14
Swamp and marsh	2.0	800–6,000	3,000	6.0	3–50	15	30
Lake and stream	2.0	100–1,500	400	0.8	0–0.1	0.02	0.05
Total	149		782	117.5		12.2	1827

[1] After Lieth and Whittaker (1975)

measurements of a minute and non-random sample of the whole biosphere. In addition they are based on measurements which rarely if ever represent actual net primary production. By definition, net primary production P_N is the total organic weight or energy gain through photosynthesis less all of the respiratory losses (Eqn. 3.1). Continuous measurements of the gaseous fluxes of CO_2 into and out of the plants of a community would be needed to directly determine P_N. At present this is technically a very difficult measurement and has only been used with any success in natural communities in a handful of instances (Eckhardt, 1975; see Montieth, 1976). Most commonly production has been estimated from measurements of the sum of change in plant community biomass and losses of this biomass through death and grazing. Equation 3.2 is a general equation for this determination adopted by IBP (the International Biological Programme, e.g. Newbould, 1967; Milner and Hughes, 1968).

$$P_N = \Delta B + L_d + G \quad \ldots\ldots\ldots\ldots (3.2)$$

where

ΔB = Biomass change over a specified time interval (Δt)
L_d = Plant losses by death and shedding during interval (Δt)
G = Direct plant losses to consumer organisms such as herbivores and parasites during interval Δt.

Whilst this method appears relatively simple at first sight, close analysis raises many problems. First, B can be measured by harvesting sub-samples of vegetation in natural ecosystems. However, heterogeneity results in large variation between samples and decreased precision. Even when a large number (n>50) of replicates are used in a simple grassland community confidence limits on estimates of mean B in the region of ±40% are not uncommon (Singh et al., 1975). The second problem lies in the estimation of unseen biomass. Only a small proportion of productivity studies take any account of the below-ground biomass yet this may, for example, amount to anything from 24%, in oak (*Quercus* sp.) forest, to 83%, in dwarf shrub tundra, of the total biomass (Rodin and Basilevic, 1966). A further problem is the exudation of organic compounds from roots into the soil and leaching into water. The few available estimates suggest that this could account for up to 50% of P_N (Bowen, 1980). The total amounts of biomass lost through disease, senescence and grazing is even more difficult to measure. Various techniques have been devised for estimating L_d, but most can be expected to underestimate its true value (Singh et al., 1975).

Different procedures for estimating production through equation 3.2 have been used for different ecosystems. With respect to methods of estimating terrestrial production two broad groupings of ecosystems can be made: (1) Ecosystems dominated by herbaceous plants, i.e. grasslands, scrublands, wetlands and macrophyte dominated waters and (2) Ecosystems dominated by trees. The specific limitations to production estimates for these two ecosystem groupings will be considered in turn.

3.2.4 Grassland, Scrubland and Wetland Ecosystems

The International Biological Programme (IBP), the basic source for the summaries in Table 3.5, has provided a considerable quantity of data for these ecosystems (Cooper, 1975).

The majority of production estimates available for this group of communities derive from direct measurements of biomass. The biomass of herbaceous plants may be harvested easily and their dry weights measured. Two problems do, however, arise. Biomass refers strictly to organic weight and many studies have not taken account of the inorganic or ash component of the dried material. Ash contents for vegetation typically range from 10–30% of the dry weight, amounting to a very significant source of error (Jørgensen, 1979). In IBP, a reduction of 5% from dry weight values was suggested to account for this error (Newbould, 1967).

Biomass, by definition, refers to living material. It is a simple if tedious matter to separate healthy green leaves from rotting dead ones in harvested material; it is even more difficult to decide how to treat senescing leaves which are often indistinguishable from dead leaves and still more difficult to deal with organs which show sequential senescence such as grass leaves which may have dead tips and healthy bases.

The P_N estimates used for IBP were largely based on peak biomass of individual species (Cooper, 1975). Indeed the majority of P_N estimates for these ecosystems are based on peak biomasses or dry weights for either the whole community or individual species (Eqn. 3.3). Sometimes this is elaborated to take account of the fact that some biomass is present throughout the year and thus it is only the difference between the temporal maximum and temporal minimum that may be equated to production (Eqn. 3.4). Thus either:

$$P_N = B_{max} \quad\quad\quad\quad\quad\quad\quad\quad (3.3)$$

or

$$P_N = B_{max} - B_{min} \quad\quad\quad\quad\quad\quad\quad\quad (3.4)$$

where:

B_{max} and B_{min} = the maximum biomass and minimum biomass of a species or community attained, respectively, in a 12 month period.

The assumption underlying these methods is that these figures give an approximate answer to Eqn. 3.2.

Many ecosystems do not have a 12 month growing season because for part of the year drought and/or temperature prevent growth. The vegetation goes through a phase of comparatively rapid production in the growing season to reach a peak towards the end of that season. Equations 3.3 and 3.4 can only be valid if losses (L_d and G, Eqn. 3.2) are zero before the peak biomass is attained and if production is zero afterwards. Any deviation would cause underestimation of P_N. This was recognized by the contributors to IBP (Cooper, 1975), but by how much is P_N underestimated?

L_d and G may of course be measured. By the use of paired plots, one cleared of dead vegetation, accumulation of dead material over a subsequent period may be estimated (Coombs and Hall, 1982). Alternatively, change in the amounts of litter with time can be recorded. However, the litter is continually decomposing and so change in the amount of litter will be an underestimate of loss through death and shedding (L_d). Decomposition of dead material has been estimated from dry weight decreases of marked leaves, dry weight losses of samples placed in fine-mesh bags (termed litter-bags), losses of radioactivity from leaves fed a radioactive substance before death, or rates of decomposer respiration (Chapman, 1976). All of these techniques suffer from certain sources of error. Litter bags, which represent the most commonly used method, can only give a true estimate of decomposition if the micro-climate of the material is not altered and the litter is a truly random sample of the litter existing at that point in time. Some alteration of micro-climate is unavoidable and commonly only a single litter sample is taken to estimate decomposition for the whole year (e.g. Wiegert and Evans, 1964; Mason and Bryant, 1973).

Grazing losses to large herbivores may be estimated by the use of exclosures. Insect grazers, especially sap sucking insects, can only be excluded by the use of very fine mesh materials which significantly alter the plant micro-climate. The little data available on seasonal changes in L_d and G, suggest that the use of equations 3.3 and 3.4 will result in serious underestimation of P_N and by a factor of as much as four times in one comparative study of techniques (Linthurst and Reimold, 1978). It is quite obvious that plants which show sequential senescence, in particular grasses, will lose large amounts of leaf material before the maximum biomass is obtained and thus serious underestimation of P_N by equations 3.3, 3.4 and related methods is inevitable.

The problems in obtaining actual P_N for this group of ecosystems can be illustrated by taking one ecosystem as a case study. Some 200 estimates of P_N have been published for temperate salt marsh communities (Turner, 1976). Of this total, 190 base their estimates of P_N purely on changes in biomass (ΔB). Some equate B to the maximum biomass recorded through one year (Eqn. 3.3), whilst others equate ΔB to the difference between the minimum and maximum recorded biomasses for one year (Eqn. 3.4). A further problem is that some studies define biomass correctly as the mass of live material alone, others define biomass as the sum of the masses of live and dead material. Only 10 studies attempt to measure losses of material through death and only 3 take any account of below-ground biomass. When the different techniques of estimating P_N are applied to the same area of salt marsh, estimates of P_N vary by a factor of 4 (Linthurst and Reimold, 1978). Thus, what appears at first sight to be a well studied ecosystem with respect to net primary production, is in fact very poorly understood.

Since it is only the above-ground biomass of standing vegetation which can usually be harvested and therefore of potential value to man, it is arguable that net primary production is of little interest. This is true where we are only concerned with harvestable biomass, but not true where we wish to understand photosynthetic efficiency of light energy conversion into biomass in natural communities. This can only be determined by a knowledge of P_N. It is arguable that most if not all estimates of P_N for terrestrial ecosystems are gross under-estimates of the true P_N, and that the quoted efficiencies of light energy conversion into biomass are not a true measure of the photosynthetic efficiency since no account is taken of the biomass formed below-ground or that lost by any cause before harvesting.

Reported estimates of production of roots and other below-ground organs must, due to the greater difficulties of measurement, be even more limited in accuracy. Unlike above-ground material the actual extraction of roots is a difficult procedure. Trenches must be cut or cores taken, the latter though quicker becomes difficult on stony soils and where root systems penetrate into a hard or rocky sub-soil. Roots may be washed out of the soil sample, but separation becomes very difficult if the soil has a high humus content and inevitably, some of the fine roots will be lost. In the IBP it was accepted that only the major roots and below-ground storage organs could be extracted in most instances (Newbould, 1967). Where detailed analyses of below-ground biomass have been conducted it may be seen that the IBP procedure underestimated below-ground biomass by 50% (Dunn, 1981; Hussey and Long,

1982). A further problem is the separation of live below-ground material. Live and dead roots are often visually indistinguishable and in wetland ecosystems both dead and old, but living, roots may be coated with black sulphide deposits. Vital staining is the most frequent method used to identify live roots, the tetrazolium dyes being commonly used for this procedure (Jacques and Schwass, 1956). Flotation methods have been suggested to supply a less laborious method of separating live and dead roots, but the effectiveness of the method is apparently very species dependent (Chapman, 1976). Hand sorting of live stained material appears the only method to provide reliable and consistent results, but it is extremely time-consuming. Even this method may result in underestimation of the fine root fraction by 40% (Hussey and Long, 1982). Production below-ground was estimated in IBP and commonly elsewhere as the difference between the maximum and minimum below-ground biomass (Eqn. 3.4; Dahlman and Kucera, 1965). As with above-ground biomass P_N is again underestimated since root mortality and disappearance during the season are not included in the production values (Cooper, 1975). Dunn (1981) in a detailed analysis of production in a salt-marsh ecosystem showed that use of eqn. 3.4 would have resulted in an underestimation of below-ground production amounting to roughly 75%.

It is impossible to accurately quantify the likely systematic errors in the IBP data which represent the most extensive and important information available for this group of ecosystems. The errors result primarily from a failure to take full account of continual turnover of biomass through the year. From the little comparative data available this underestimation could be as much as 75% for above-ground production and even more for below-ground. Better estimates await more detailed measurement of production which include measurements of losses through death and shedding, through grazing and through root exudation. This applies particularly to tropical ecosystems, for which there is little comparative methodological data and where rates of turnover of living material and decomposition of dead material would probably be larger than for the studies summarized above.

Finally, the influence of year by year variation in production of these ecosystems should not be overlooked. Many published studies have extended only over a single year. Jackson (unpublished data) in a detailed study of a coastal salt marsh in Suffolk, England, found that production varied by ±20% from year to year. This variation must certainly be greater in the semi-arid tropics and tundra where year to year variation in rainfall and temperature, respectively, will have the most profound influence on production.

3.2.5 Forest Ecosystems

The problems and limitations of production measurements in ecosystems dominated by herbaceous plants apply equally to the leaves and small roots of trees. In IBP it was recommended to measure root production from seasonal change in biomass (Eqn. 3.4, Newbould, 1967). Leaves, are of course, far more difficult to harvest from trees than from herbaceous plants. However, in deciduous forests with a single phase of leaf growth a good estimate of leaf production may be obtained, providing grazing has been insignificant, simply by collecting leaves at the time of shedding (Newbould, 1967; Chapman, 1976). For evergreen species the difference between minimum and maximum leaf biomass was suggested as an estimate of leaf production in IBP (Newbould, 1967), but with the attendant problems already outlined for herbaceous species. Tropical rainforests are evergreen with a tremendous range of leaf forms and patterns of leaf growth, senescence and decomposition. Errors in estimates of leaf production for this ecosystem will most certainly be greater than for any other. Stems are suggested to represent the bulk of production in woody plants. Since the stems and major roots of trees are perennial, living for many years, turnover of material is far slower than in herbaceous communities. In mixed Oak-pine forests the ratio of production to biomass was estimated as 0.08 (Jørgensen, 1979) compared to >1 for herbaceous communities. For an individual tree, production of woody material may be closely estimated from change in biomass since losses by death and grazing will be small until the whole tree dies. Thus for production of wood:

$$P_N = \Delta B \quad \quad (3.5)$$

The major problem is that it is simply not practicable to harvest any number of trees or large shrubs and measure their dry weight. Instead annual production of wood may be estimated from the width of annual growth rings measured from cores cut out of the trunk (Newbould, 1967).

Stem biomass is often estimated from regression equations (based on the harvest of a few plants) of biomass against a more easily measurable parameter such as height or most commonly the stem diameter at breast height (DBH). For major timber trees detailed regressions or yield tables relating biomass to DBH are available. This approach will provide accurate estimates of stem biomass in woodlands composed of trees of similar form, and preferably where harvests have been made at regular intervals. Clearly the more variable the form and age of the trees and the more species diverse the forest, the greater the difficulty of obtaining an accurate regression relationship and the greater the number of regressions required. Thus, the precision of estimates obtained for the same amount of effort will be much smaller in tropical rainforests which have a much greater range of species and absolute sizes of trees than in temperate and boreal forests which contain fewer species spanning a smaller range of heights. This difference in precision is very significant on a world scale since tropical forests are suggested to account for more than half of the world's total plant biomass and more than a quarter of world primary production. Extrapolating from detailed regressions of stem-wood production against DBH for mixed woodlands in the U.S.A. the variation about the regression for trees of 2m DBH, for example, ranges from about 600 g yr^{-1} to 3000 g yr^{-1} or an error on the regression estimate of ±60% (Whittaker and Woodwell, 1968).

3.3 AGRICULTURAL SYSTEMS

In contrast to natural ecosystems, agricultural systems are deliberately modified by man. It should be remembered that the economic viability of a crop or cultivation practice is dependent on the productivity obtained. In contrast to natural ecosystems, many detailed and precise estimates of the productivity of crops exist. However, interest has naturally focused on the harvested material and as with natural ecosystems, few studies have been concerned with measurement of production of roots and rhizomes, or parts of the plant which die and are shed before harvesting.

Estimates of net primary production from cultivated land vary from 1 to 88 t ha^{-1} and reflect the confounding of environment, cultivation practice and economic restraints on the expression of genotypic or environmental potential of crop plants (Table 3.6; Loomis and Gerakis, 1975; Buringh, 1980). For many agricultural or horticultural species only a proportion of the plant is of economic importance and

Table 3.6 Good yields of dry matter production[1]

	Type	Annual yield t ha^{-1} yr^{-1}	Growth rate g m^{-2} d^{-1}	Conversion efficiency % total radiation
Sub-tropical/tropical				
Pennisetum purpureum	C$_4$	88	24	1.6
Saccharum officinarum	C$_4$	66	18	1.2
Zea mays	C$_4$	27	23	0.8
Pennisetum typhoides	C$_4$	21	19	—
Sorghum spp.	C$_4$	28	23	—
Oryza sativa	C$_3$	22	—	—
Manihot esculentra	C$_3$	33–41	11	0.5
Elaeis guineensis	C$_3$	29	—	—
Medicago sativa	C$_3$	30	8.1	—
Glycine max	C$_3$	9	—	—
Annual crops	—	30	6.8	—
Perennial crops	—	75–80	—	—
Rain forest	—	35–80	—	—
Temperate				
Beta vulgaris	C$_3$	22–34	12–14	—
Triticum aestivum	C$_3$	18–30	—	—
Solanum tuberosum	C$_3$	22	—	—
Lolium perenne	C$_3$	22	—	—
Perennial crops	—	29	—	1.0
Annual crops	—	22	6	0.8
Grassland	—	22	6	0.8

[1] See Boardman (1977, 1978), Hall (1979), Loomis and Gerakis (1975), for original references.

Table 3.7 Some high, short-term, dry weight yields of crops and their short-term photosynthetic efficiencies[1]

Crop	Type	Country	Yield (g m^{-2} d^{-1})	Efficiency (%)
Temperate				
Festuca arundinacea	C$_3$	U.K.	43	3.5
Lolium perenne	C$_3$	U.K.	28	2.5
Dactylis glomerata	C$_3$	U.K.	40	3.3
Beta vulgaris	C$_3$	U.K.	31	4.3
Brassica oleracea	C$_3$	U.K.	21	2.2
Hordeum sativum	C$_3$	U.K.	23	1.8
Zea mays	C$_4$	U.K.	24	3.4
Triticum aestivum	C$_3$	Netherlands	18	1.7
Pistum sativum	C$_3$	Netherlands	20	1.9
Trifolium pratense	C$_3$	New Zealand	23	1.9
Zea mays	C$_4$	New Zealand	29	2.7
Zea mays	C$_4$	U.S., Kentucky	40	3.4
Sub-tropical				
Medicago sativa	C$_3$	U.S., California	23	1.4
Solanum tuberosum	C$_3$	U.S., California	37	2.3
Pinus spp.	C$_3$	Australia	41	2.7
Gossypium hirsutum	C$_3$	U.S., Georgia	27	2.1
Oryza sativa	C$_3$	Australia	23	1.4
Saccharum officinarum	C$_4$	U.S., Texas	31	2.8
Sorghum sudanense	C$_4$	U.S., California	51	3.0
Zea mays	C$_4$	U.S., California	52	2.9
Tropical				
Manihot esculenta	C$_3$	Mayalsia	18	2.0
Oryza sativa	C$_3$	Tanzania	17	1.7
Oryza sativa	C$_3$	Philippines	27	2.9
Elaeis guineensis	C$_3$	Malaysia (whole year)	11	1.4
Pennisetum purpureum	C$_4$	El Salvador	39	4.2
Pennisetum typhoides	C$_4$	Australia	54	4.3
Saccharm officinarum	C$_4$	Hawaii	37	3.8
Zea mays	C$_4$	Thailand	31	2.7

[1] From Hall (1979). These figures reflect seasonal and growth-related factors which result in faster short-term rates of photosynthesis than are observed when data are calculated on an annual basis.

in several instances they have been deliberately bred to optimize economic yield at the expense of biomass yield. Estimates of production between regions vary from 1 to 28 t ha^{-1} yr^{-1} of cereal grain and 1 to 88 t ha^{-1} yr^{-1} of harvestable dry matter for forage crops, excluding their roots (Buringh, 1980). Maximum economic yields of agricultural systems are often achieved at planting densities which are less than those for maximum primary production (Loomis and Gerakis, 1975).

Although hundreds of species have been domesticated, 90% of the world's food is provided by just 24 species. A further group of species, mainly grasses, have been domesticated as fodder for man's domestic animals. The number of varieties or cultivars of these species can be quite large as the result of intensive breeding programmes which have been pursued to enhance yield and improve disease resistance. A useful approach for assessing productivity in these agricultural crops is to collate estimates of good annual yields (Table 3.6). Maximum annual yields

Table 3.8 Actual production and demand for wheat, rice and coarse grains[1] (kt)

1972-74	Production Developing	Production Developed	Demand Developing	Demand Developed
Wheat	108 295	250 321	141 026	214 095
Paddy rice	299 427	24 051	304 117	22 821
Coarse grains	213 963	440 306	214 117	438 404

[1] From FAO Agricultural Commodity Projections, 1975–85.

Table 3.9 Share of world production (%) of wheat, rice and maize by the principal nations involved in growing these crops.[1]

	Wheat	Rice	Maize
Argentina	—	—	3.2
Bangladesh	—	5.9	—
Brazil	—	2.0	4.8
Canada	4.5	—	—
China	7.6	33.1	8.0
France	4.9	—	3.4
India	6.7	21.1	—
Indonesia	—	7.3	—
Italy	—	3.3	—
Mexico	—	—	3.9
Japan	—	4.9	—
Others	29.3	10.7	18.2
Pakistan	2.1	—	—
Rumania	—	—	2.2
South Africa	—	—	3.5
South Korea	—	1.9	—
Sri Lanka	—	4.6	—
Thailand	—	3.9	—
Turkey	2.2	—	—
U.S.S.R.	30.0	—	4.2
U.S.A.	12.7	1.3	46.0
Yugoslavia	—	—	2.6

[1] From Loomis and Gerakis (1975).

are realized only with maximum inputs of nutrients, a good supply of water, optimum climatic conditions and pest control. Such a combination is unlikely to be feasible particularly in countries where economic constraints dominate productivity or in developed countries where net margins are commensurate with economically optimum, rather than maximum, yields.

An analysis of the overall photosynthetic efficiencies of plants shows that they are normally less than 1% in temperate species, which are usually C_3 and only exceed 1% in tropical C_4 species (Table 3.6). This is clearly a consequence of the higher rates of photosynthesis of C_4 plants, particularly at high light intensities, and their ability to suppress photorespiration. C_4 plants include many important commercial crop plants such as *Zea mays, Sorghum bicolor, Saccharum* spp. and various species of millet, as well as high yielding forage grasses, such as *Pennisetum purpureum,* (Table 3.7). In general, therefore, maximum biomass production by C_4 plants exceeds that of terrestrial C_3 plants (Monteith, 1978; Osmond *et al.*, 1982). There is however a strong correlation between latitude and performance of plants. C_4 are superior to C_3 species at low but not at high latitudes, the cross-over point being related to radiation level and the occurrence of chilling temperature (Loomis and Gerakis, 1975). These differences are substantiated by measurements of short term crop growth rates (Table 3.7). Maximum values are equivalent to photosynthetic energy conversion efficiencies exceeding 4%. [Unusual performance by some species (6%) that exceed theoretical predictions are thought to arise partially from the use of small plots which receive more radiation than a plot in a uniform stand (Loomis and Gerakis, 1975).]

World food production currently exceeds that required by the world population by 10% and countries in food surplus could feed those in food deficit if political and economic constraints were removed (Hall, 1983). This has not occurred and the demand from developing countries for human food continues. Their continuing food deficits and the recent interest in biomass production for energy emphasize the need to improve average yields in each individual country. The total production and demand for wheat, rice and coarse grains are given in Table 3.8. This clearly points out that for each major category production exceeds demand in developed countries while the reverse is the case in developing countries. FAO commodity projections (1979) suggest that this situation will persist. Even with sustained increases in production at current rates of expansion and the most optimistic assumptions, there is no prospect of the gap narrowing. The share of production between countries of wheat, rice and maize is given in Table 3.9. The major centres for wheat are the U.S.S.R. and the U.S.A.; for rice China, India, and Indonesia, while production of maize is dominated by the U.S.A.

The dramatic increase in the yield of crops in developed countries has materialised from a combination of plant breeding and substantial inputs of high cost technology. There is some evidence that mean yields are levelling off though these may be substantially lower than the maximum yield, even in developed countries (Jensen, 1978). Substantial increases in production could, therefore, be realised from increasing the productivity of existing cultivated land.

3.4 BIOMASS

The current interest in the production of "biomass" relates to the harvest of plants as a source of fuel (Cote, 1983). This has become of increasing importance in several contexts. First, half the trees felled at present in the world are used directly as fuelwood for cooking and heating particularly in developing countries, but are not being replaced to ensure a continuing supply. Secondly, biomass production is being given serious consideration in developed countries as an alternative and renewable source of energy. In both instances the need for biomass production has arisen because of the problem of financing oil imports or the availability of adequate supplies of fossil fuel.

The major emphasis has been on food crops with a high carbohydrate content for liquid fuel from yeast-based ethanol fermentation, and on fast-growing tree crops or agricultural wastes with a high lignin/cellulose content for solid fuel for direct combustion or feedstock for pyrolysis (Hall et al., 1982; Holdgate et al., 1982). More recently vegetable oils have been used as substitute diesel fuel.

In contrast to natural communities where the productivity of the system can be seen in ecological terms and in agricultural systems where productivity includes parameters of taste, quality, nutritive value and economic yield, the major requirements in biomass production are quantity and energy value i.e. $GJ\ m^{-2}\ yr^{-1}$ (Coombs et al., 1983). The net energy yield will be the difference between the energy content of the biomass which can be harvested and the energy which has been put into the system in order to produce and harvest the biomass. In most instances biomass crops will be plants grown as monocultures e.g. *Saccharum officinarum* and *Salix* spp., and to some extent maximum biomass yields will be commensurate with planting densities which are higher than those used in agricultural/forestry systems under similar environmental conditions. With determinate crops total dry matter yields per unit area, unlike economic yield, will tend towards a constant value independent of planting density.

Part II
Factors Influencing Photosynthetic Productivity

Introduction

Photosynthesis is a physical and chemical process requiring an atmosphere of carbon dioxide, water and light. Green plants are subjected to an environment which modifies the rate of these physical and chemical processes and under certain conditions causes them to reversibly or irreversibly cease. During the day, light fluctuates sufficiently to limit both single leaf and crop photosynthesis. In addition, concern is now being expressed about the continued increase in the atmospheric CO_2 concentration which increases photosynthesis and may favour certain species in preference to others.

Environmental factors which affect photosynthetic productivity are related to solar radiation, the weather, edaphic factors and pollutants. This section considers each in turn and their general effects are summarized in Table 4.1. This summary intends only to indicate general trends, and individual exceptions to these trends can be found for most combinations of environmental effect and photosynthetic mechanism.

CHAPTER 4
Light

4.1 AVAILABLE LIGHT

Sunlight energy for biomass production is obtained from the total short wave radiation incident at the earth's surface. The maximum quantity of direct sunlight incident on plants at sea level is 900 W m^{-2} (Gates, 1965) but as the total quantity of light available to plants fluctuates with latitude and time of year, the mean global irradiance fluctuates over the surface of the earth (Fig. 4.1). Variations in the temporal distribution of solar irradiance also occur on a daily basis throughout the growing season. The major proportion of this radiation (99%) is in the waveband 0.3–4 µm. The photosynthetic pigments of terrestrial plants use light from the visible spectrum only (0.4–0.7 µm) which is also referred to as photosynthetically active radiation (PAR, Ludlow, 1982). Part of this PAR is wasted due to the physical properties of leaves e.g. by reflectance and transmission and through fundamental thermodynamic considerations which limit the conversion and storage of sunlight as chemical energy in photosynthesis (Bunnik, 1978; Good and Bell, 1980).

Light is obviously fundamental to photosynthesis and the basis of its and our existence, but at the same time a major factor limiting biomass production. Some light is available throughout the year in all but polar latitudes, but other constraints, particularly temperature and water, define growing seasons rendering net photosynthesis impossible in spite of the presence of adequate light. The potential conversion efficiency of photosynthesis is thus further reduced. Of the light remaining, that intercepted by chlorophyll is the discriminant of biomass production, not the light incident above the crop, whilst the length of the growing season determines the maximum quantity of light which can be intercepted for photosynthesis (Monteith, 1981).

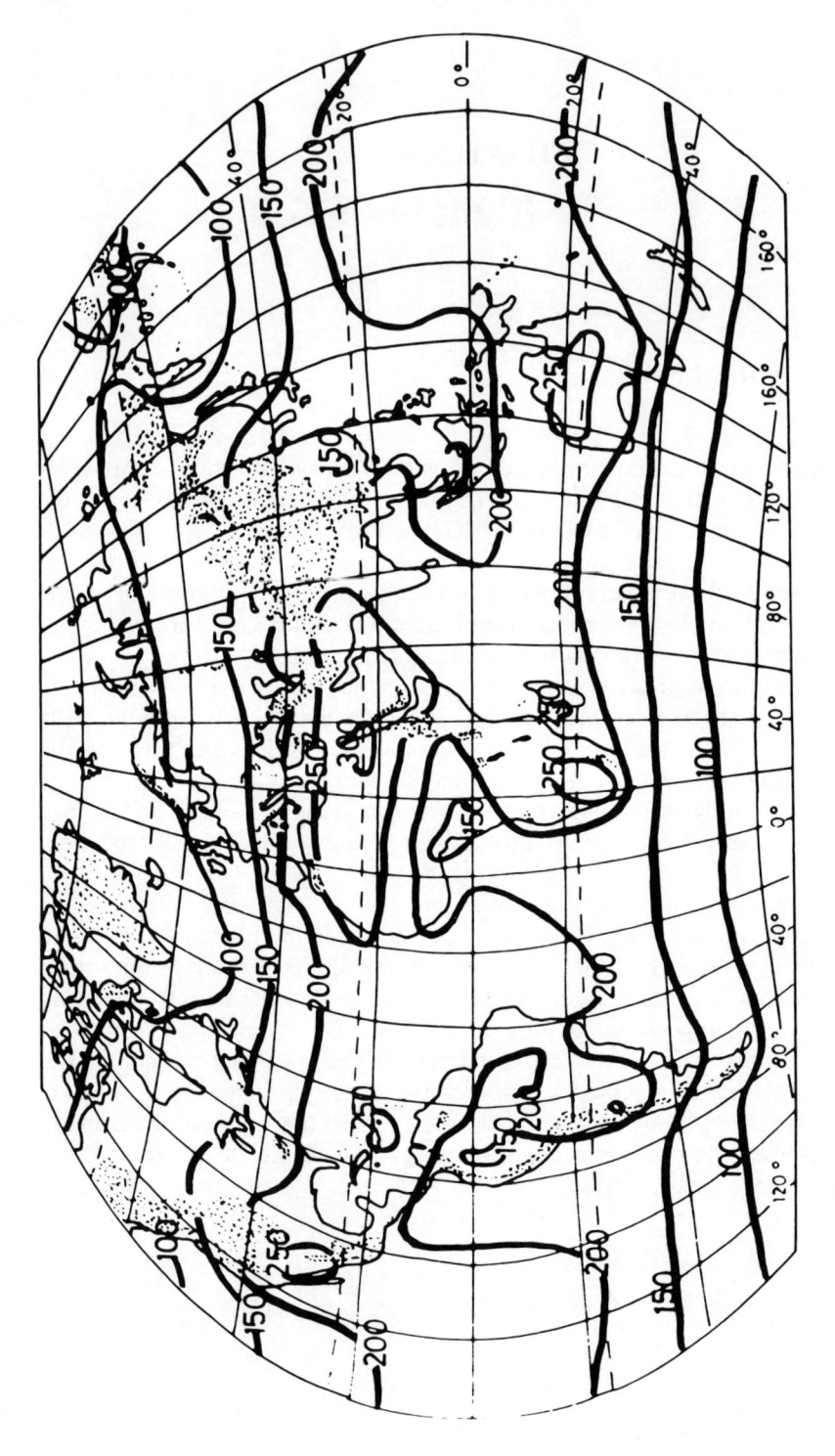

PHOTOSYNTHESIS IN RELATION TO PLANT PRODUCTION

Fig. 4.1 Annual mean global irradiance on a horizontal plane at the surface of the earth (W m^{-2} averaged over 24 hours, reproduced with permission from Coombs & Hall, 1982).

Table 4.1 Summary of Effects on Environmental Changes on different components of the photosynthetic process in Crops.

	1. Increased CO_2					6. Salinity				
	2. O_3 Pollution					7. Water Stress				
	3. SO_2 Pollution					8. Low Temperature				
	4. Nitrogen Stress					9. High Temperature				
	5. High Nitrogen					10. Light				

	1.	2.	3.	4.	5.	6.	7.	8.	9.	10.
Photochemistry	0	?	?	--	?	+	-	0	0	++
Electron transport	0	-	-	?	?	+	-	-	+	++
Carbon metabolism	++	?	-	?	-	?	-	--	+	+
Glycolate synthesis	--	?	?	?	?	?	++	--	++	+
Stomatal conductance	--	-	-	-	?	-	--	--	--	0
Leaf photosynthetic rate	+	-	-	-	-	-	--	--	-	+
Dark respiration	0	+	+	-	?	+	--	--	++	0
Leaf area development	+	?	?	--	++	--	--	--	-	+
Leaf death	0	+	+	++	-	++	++	-	++	-
Crop photosynthetic rate	+	-	-	--	++	--	--	--	--	++
Productivity	+	-	-	--	++	--	--	--	--	++

++	Marked increase	-	Occasional or non-linear decrease
+	Occasional or non-linear increase	--	Marked decrease
0	No obvious change	?	Response uncertain

4.2 CONVERSION OF ENERGY

It is common practice to express the conservation of energy by crops as the efficiency of use of sunlight (energy conserved in biomass/energy content of sunlight incident on crop). Published figures of 1% are used to justify an obvious need to seek better conversion efficiencies but it is important to put potential improvements into perspective (Loomis and Williams, 1963; Yocum et al., 1964). The sequence of light harvesting and electron transport which incorporates the energy of sunlight into ATP and NADPH through the splitting of water molecules can be summarized as,

$$2H_2O + \text{sunlight} \longrightarrow O_2 + 4H^+ + 4e^- \quad \ldots\ldots (4.1)$$

Carbon dioxide is then reduced to simple carbohydrate, so it follows that,

$$4H^+ + 4e^- + CO_2 \longrightarrow (CH_2O) + H_2O \quad \ldots\ldots (4.2)$$

As may be seen from the second equation, four electrons must be transferred to reduce 1 molecule of CO_2 to carbohydrate but as two photosystems are utilized to transfer electrons from water to NADPH, a minimum of eight photons must be absorbed. If all PAR was usefully absorbed by the leaf, it may be calculated from the energetics of the reactions in equations 4.1 and 4.2 and assuming the minimum quantum requirement for photosynthesis of eight photons, that the free energy stored per mole of CO_2 reduced is 27–28% (UK-ISES, 1976; Bassham, 1977; Good and Bell, 1980). The inactive absorption, spectral reflectance and transmission of PAR by crops is a complex function of canopy and leaf structure (Loomis and Williams, 1963; Bunnik, 1978). In practice minimum losses of around 10% are probably incurred and reduce the photosynthetic energy conversion of PAR to 24–25% (Yocum et al., 1964). As PAR constitutes only 50% of solar radiation (Monteith, 1973) it follows that the maximum conversion efficiency of solar radiation into photosynthetic products has an apparent upper limit not exceeding 12% (see Bolton, 1978; Good and Bell, 1980; Prioul, 1982; Varlet-Grancher et al., 1982; Beadle and Long, 1985). Rates of dark respiration in C_4 plants and rates of dark and photorespiration in C_3 plants reduce this upper limit to 3.7–4.4% and 5.0–5.8% of solar radiation in C_3 and C_4 plants respectively (Table 4.2; UK-ISES, 1976; Beadle and Long, 1985).

Table 4.2 The partition of sunlight energy from that incident on vegetation by crops.[1]

Energy losses due to:	% loss	% remaining
Energy outside the photosynthetically active waveband	50	50
Reflection and transmission	5–10	40–45
Inactive absorption	2.5	37.5–42.5
Photochemical inefficiency	8.7	28.8–33.8
Carbohydrate synthesis (max. loss in C_4 photosynthesis)[2]	18.9–22.2	9.9–11.6
Photorespiration (C_3 plants only)	2.5–2.9	7.4–8.7
Dark respiration in C_4 plants	4.9–5.8	5.0–5.8
in C_3 plants	3.7–4.3	3.7–4.4

[1] The figures given are in relative terms (total short wave radiation, i.e. solar radiation is assumed to be 100%).
[2] The conversion of excitation energy to glucose assumes that 1 mole of light at 690 nm absorbed during photochemistry contains 173.3 kJ. As 1 mole of carbohydrate conserves 477.0 kJ, it follows that the efficiency (if $\phi = 8$) is $477/(8 \times 173.3)$ or 34.4% of excitation energy.

A value approaching this theoretical upper limit has been observed for *Zea mays*, a C_4 species, during the period of maximum growth and following canopy closure (Lemon, 1965). This provides a clear indication that crop plants are able to photosynthesize at their maximum potential when other factors are non-limiting. Seasonal maxima and annual conversion efficiencies are considerably less (Bassham, 1977). It is not possible to accurately quantify the sources of these further losses, but they relate to environmental factors other than light, which include agricultural practice, pests and diseases, genetic limitations, growth patterns, assimilate partitioning and harvest yield. Even during the growing season, an early lack of total crop cover and later a lowered photosynthetic activity of older leaves, reduce light absorption and conversion efficiency. For example, 86% of the final yield of a sugar beet crop was produced in just 44% of the growing period (Gaastra, 1965).

4.3 QUANTUM YIELD

The parameter of efficiency of light utilization by photosynthesis is the quantum yield (φ), the moles CO_2 fixed per mole quanta absorbed by a leaf (mole/mole). Since light becomes of less importance as a factor limiting photosynthesis with increasing quantum flux density, Q, the true quantum yield can only be measured at low Q when photosynthesis is strictly light limited and proportional to Q. Quantum yield is a dimensionless constant which has a maximum value of 0.125 for photosynthesis i.e. from the minimum quantum requirement (1/φ = 8). Reflection, absorption by substances other than photosynthetic pigments (e.g. anthocyanins), fluorescence and radiationless decay of excited pigment molecules all reduce quantum yield. Typically these processes account for just over one quarter of the absorbed energy so that the maximum quantum yield of gross photosynthesis is reduced to 0.09 (1/φ = 11) compared to an observed quantum yield of 0.0733 ± 0.0008 (1/φ = 13.6) for several C_3 species in air of 2% oxygen which inhibits photorespiration (Ehleringer and Björkman, 1977). This difference between the measured and estimated requirement is probably due to the energy demand of other light activated processes, e.g. nitrogen and sulphur metabolism (Lea and Miflin, 1979; Schmidt, 1979).

In practice, the observed quantum yield of several C_4 plants and several C_3 plants in air of 21% oxygen were similar at 30°C, φ = 0.0524 ± 0.0014 and φ = 0.0534 ± 0.0009 respectively (Fig. 4.2a). These additional energy requirements can be predicted from the pathways of photosynthesis and photorespiration in C_3 and C_4 plants (Campbell and Black, 1978). Coincidentally, the additional energy requirements of C_4 photosynthesis are approximately offset by those of photorespiration in C_3 photosynthesis, at 30°C (Ehleringer and Björkman, 1977). If light is a limiting factor in biomass production therefore, any differences in performance of C_4 and C_3 plants at this temperature are less related to light limiting conditions and quantum yield but more to maximum rates of photosynthesis under light saturating conditions. Differences in the competitive ability and therefore biomass production of C_4 and C_3 plants, however, may well be related to quantum yield because of the negative correlation with temperature in C_3 plants due to photorespiration (Ehleringer, 1978).

The constancy of φ within C_3 species, when measured either in 2% oxygen or saturating CO_2 concentrations (Fig. 4.2b), and within C_4 species in normal air (Ehleringer and Björkman, 1977), is consistent with the theoretical explanation that φ should be similar amongst plants that use identical photosynthetic mechanisms in a physicochemical process which is independent of temperature – when light is the only limiting factor (Björkman, 1981). The position of C_3 plants which photorespire in air is different and there are marked changes of φ with temperature compared to C_4 plants where φ remains constant with temperature (Ehleringer and Björkmann, 1977). A remaining point of contention is the existence of a unique crossover point in the value of φ between C_4 and C_3 plants, which above and below 30°C (or more possibly nearer 22°C as 30°C may have been an overestimate), confers advantages on C_4 and C_3 plants, respectively (Ehleringer, 1978; Berry and Raison, 1981). Using the results of his own comparison between C_4 grasses and C_3 legumes and those of Bull (1969), Ludlow (1980) has questioned the validity of this analysis since ɸ of C_4 species was always higher than ɸ of C_3 species, except at the extremes of temperature. A much greater energy demand for light-activated nitrite reduction in legumes may go some way to explain these contrasting results, but whatever the ecological significance of ɸ, and more recent data favours the interpretation of Ehleringer and Björkman (1977) (see Monson et al., 1982), the remarkable constancy of ɸ within C_4 and C_3 species suggests that the photosynthetic efficiency of energy conversion may be a relatively conservative property of green plants and not subject to easy manipulation.

4.4 LIGHT RESPONSE CURVE

Net photosynthesis responds hyperbolically to quantum flux density as light becomes of decreasing importance as a limiting factor. Individual leaves of C_3 plants are typically unable to use additional light above about 500 µmol m^{-2} s^{-1}, roughly 25% of full sunlight, but this is not true of C_4 plants which in general fail to saturate even at full sunlight (Fig. 4.3).

Photosynthetic capacities of plants measured under saturating light conditions suggest that there is considerable variation both within and between C_4 and C_3 species. Maximum rates of photosynthesis of C_4 plants exceed those of C_3 plants; those of C_4 grasses (1.4–2.9 mg m^{-2} s^{-1}) are the highest recorded (Körner et al., 1979; Nobel, 1980a). The photosynthetic capacity is a function of the environmental conditions to which the plant is subjected during its growth and development but even under similar conditions there appears to be considerable variation even within a species. These apparent differences probably originate in the mesophyll though their biochemical basis is not clear (e.g. Bennet and Rook, 1978).

Besides marked variation within the maximum rates of photosynthesis, there are also exceptions to the normal relationships between net photosynthesis and light in C_4 and C_3 plants. It is not known whether these differences are directly linked to photochemical processes but clearly some plants are able to utilize their supply of light to better advantage than others and this may in some way be linked to constraints in the design of their photochemical apparatus (Woolhouse, 1978). In this respect Leverenz and Jarvis (1979) have suggested that the convexity of the light response curve of a species may be related to its productivity. This convexity should increase when the chloroplasts are more evenly illuminated and this was observed under bilateral, compared to unilateral, illumination in *Picea sitchensis*. The convexity of the response curve in this gymnosperm was also less when compared to the usually more productive C_3 angiosperms.

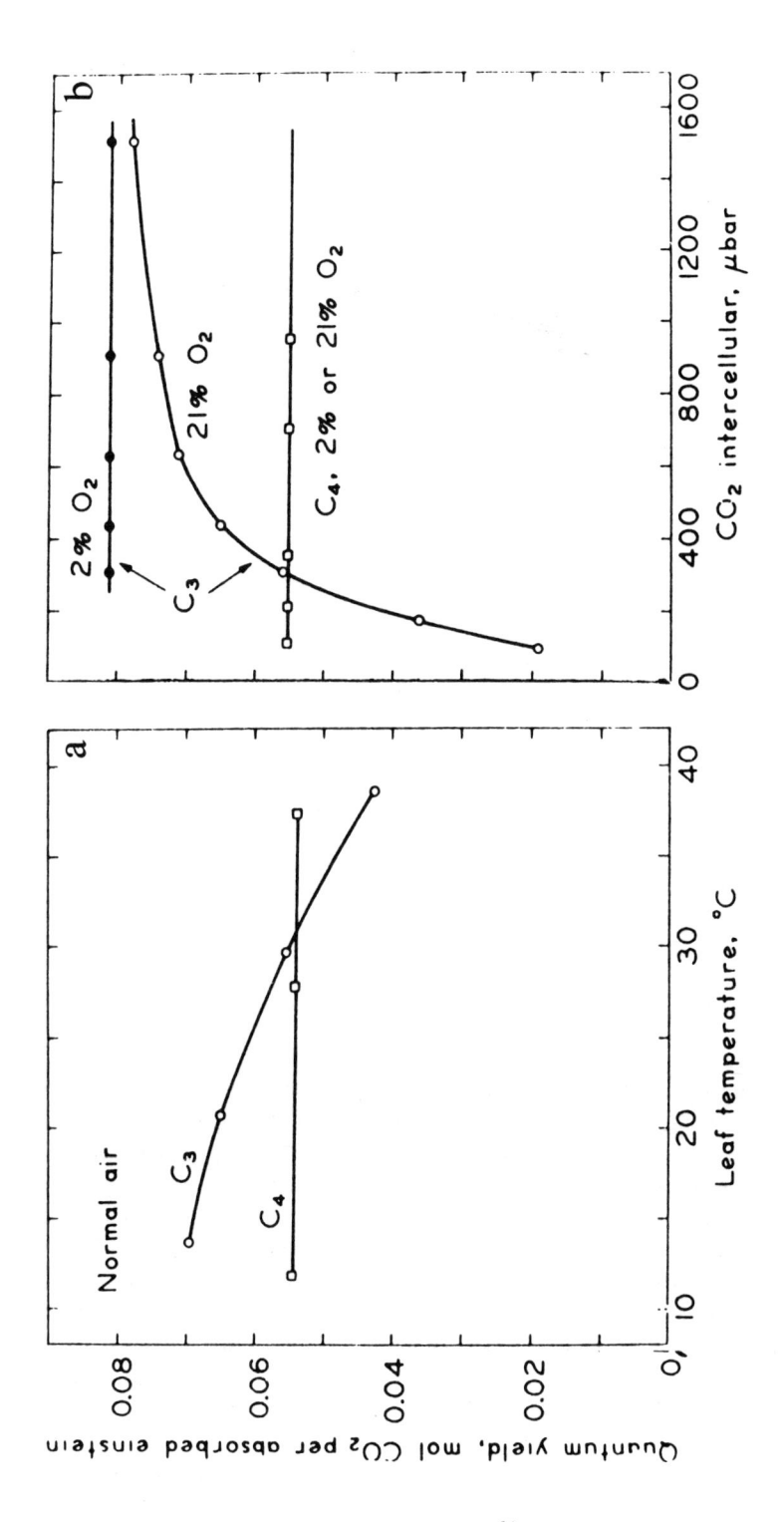

Fig. 4.2 Quantum yield for CO_2 uptake in (a) Encelia californica (C_3) and Atriplex rosea (C_4) in normal air (21% oxygen) as a function of temperature and (b) the same species in 21% or 2% oxygen as a function of intercellular CO_2 concentration. Leaf temperature was 30°C (reproduced with permission from Björkman, 1981).

Fig. 4.3 A comparison of the responses of net photosynthesis (F) for leaves of the C_3 grass Lolium perenne (open squares) with the C_4 grasses Spartina anglica (closed triangles) and Zea mays (open circles). All measurements were made at leaf temperatures of 25°C. (reproduced with permission from Long and Woolhouse, 1978).

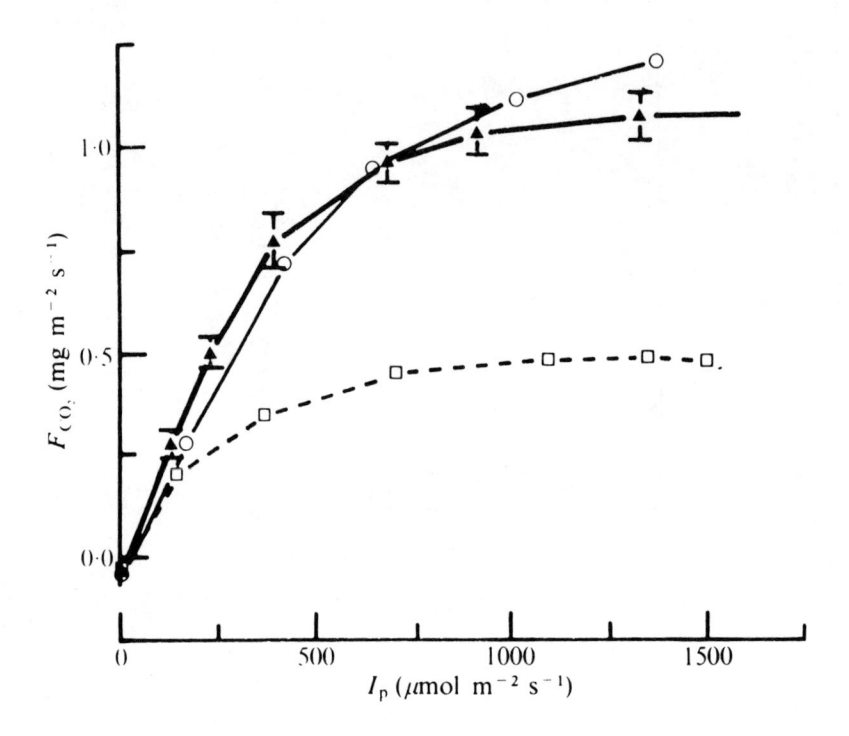

4.5 SUN AND SHADE SPECIES

Plants are divided into sun and shade species (Bohning and Burnside, 1956) though the latter are of little significance to bioproductivity since they are incapable of photosynthesis at high quantum flux densites (Fig. 4.4a). The ability of sun species to adapt to light intensity is a basic growth response (Fig. 4.4b) and these species (including most cultivated crops) integrate and adjust several partial processes to maximize photosynthesis to the available quantum flux density (Q), but with the constraint that a high photosynthetic efficiency of light utilization at one extreme of Q precludes a high efficiency at the other (Boardman, 1977). The quantum efficiency is however constant for photosynthesis irrespective of the flux density for growth (Björkman et al., 1972). In contrast, photosynthesis in shade grown plants was light saturated at *ca.* 200 µmol m^{-2} s^{-1} less, and the light compensation point was *ca.* 20 µmol m^{-2} s^{-1} less, than the levels required for the same species grown in full sunlight (Burnside and Böhning, 1957). Plants are able to adjust to changing

Fig. 4.4 Rate of net photosynthesis as a function of incident quantum flux density for (a) sun species Encelia californica and Nerium oleander and shade species Cordyline rubra and (b) for sun species Atriplex triangularis grown under three different light intensity regimes. All measurements were made in air of normal CO_2 content and at a leaf temperature of 25 or 30°C (reproduced with permission from Björkman, 1981).

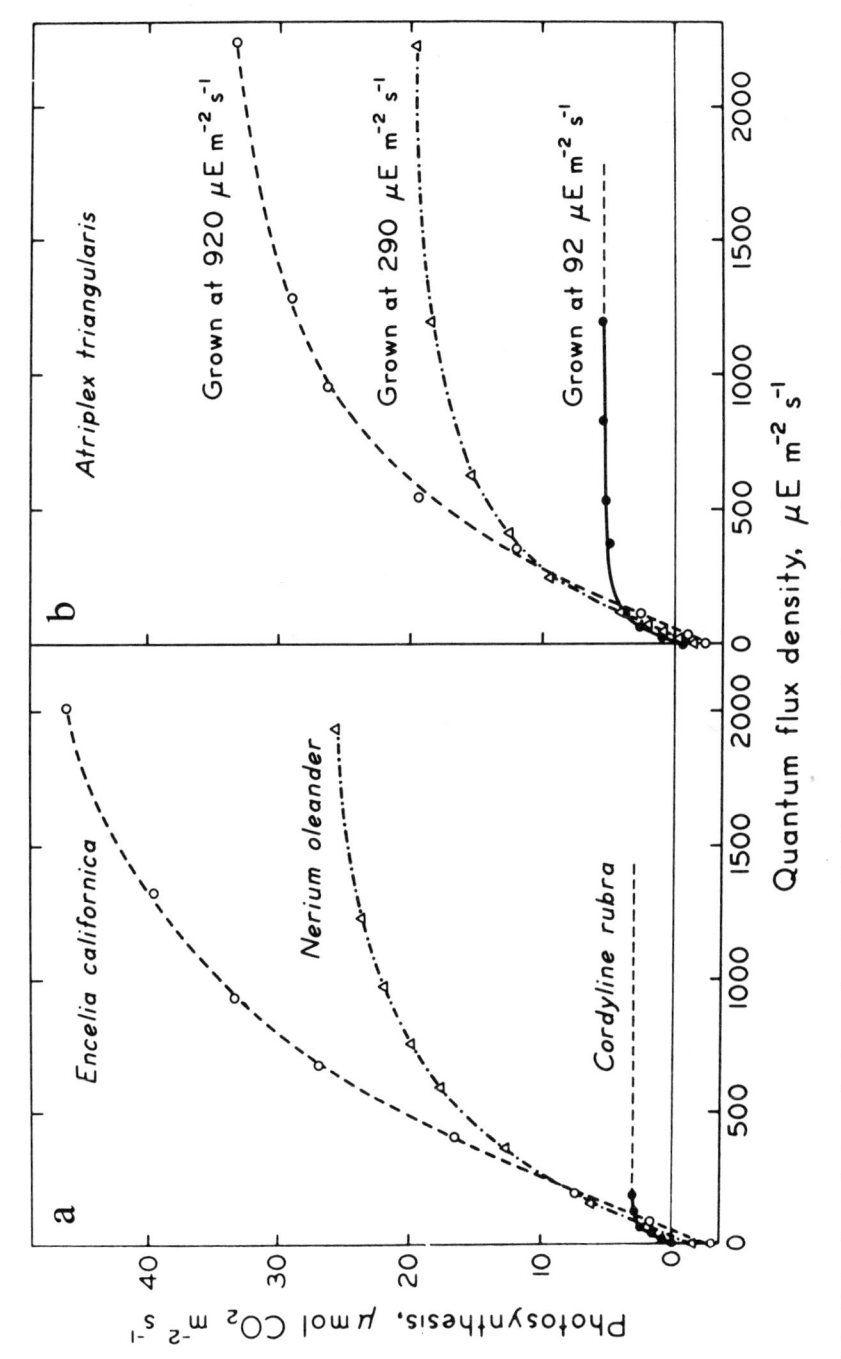

ambient light within days. For example, the photosynthetic rates of *Zea mays* grown in low light rose to that of plants kept continuously at high light six days after being transferred to the high light environment (Hatch *et al.*, 1969). Published results suggest that the capacities of C_4 and C_3 plants for light acclimation are similar (Björkman, 1981).

Major differences in photosynthetic activity in sun and shade adapted leaves are correlated with differences in the concentration of components of the electron transport chain (Björkman *et al.*, 1972; Boardman *et al.*, 1972), photosystem activity (Grahl and Wild, 1975), as well as the activity of enzymes, particularly Ru*bis*CO in C_3 plants (Björkman, 1981) and enzymes specific to C_4 photosynthesis (Hatch *et al.*, 1969). In view of the coupling of several enzymes of the carbon reduction cyles to the electron transport chain, this is perhaps not surprising.

CHAPTER 5
Temperature

5.1 TEMPERATURE

Although light is the driving force for photosynthesis, other environmental factors modify the rate of photosynthesis and production; temperature is often the most important (Cooper and Tainton, 1968; Kawashima, 1980). Differences in biomass production and growth between species correlate with the response of their photosynthesis to temperature (Berry and Raison, 1981). Other factors, of course, also correlate with biomass production, in particular the availability of water and nutrients (Ch. 6).

All photosynthetic reactions, with the exception of primary photochemistry, are thermochemical, being dependent on the probability of collision between reactant molecules. As temperature is a direct expression of the kinetic energy of these molecules photosynthesis will, in theory, increase with temperature. The net photosynthetic rate will be determined by respiratory losses of carbon dioxide, the effects of temperature extremes which are imparted both directly and indirectly on photosynthetic components, and interactions with other variables. The resultant shape of the response curve between photosynthesis and temperature is species dependent and characterized by a high and low temperature compensation point and an optimum temperature (Ludlow and Wilson, 1971a; Nobel et al., 1978).

A range of species from temperate and cool coastal habitats have different response curves to species from tropical and desert habitats. In general, temperate species show a rather flat-topped response to temperature with an optimum in the range 15–30°C and tropical species a pronounced but higher optimum: these are often C_3 and C_4 species, respectively (Fig. 5.1, Cooper and Tainton, 1968; Björkman et al., 1975; Long et al., 1975; Bird et al., 1977). Some C_3 species have maximum photosynthetic rates similar to those of C_4 species and a pronounced temperature optimum (e.g. *Encelia farinosa*, Ehleringer and Björkman, 1978). To some extent therefore, the shape of the response curve between photosynthesis and temperature may be a function of the maximum rate of photosynthesis. Björkman (1975) maintained that there was nothing intrinsic in the characteristics of C_4 photosynthesis which would determine the shape of the response.

Fig. 5.1 *A comparison of the temperature responses of net leaf photosynthesis (F) of the temperate C_4 grass* Spartina anglica *(closed triangles) and the tropical C_4 grass* Pennisetum purpureum *(open triangles) with two temperate C_3 grasses,* Festuca arundinacea *(open squares) and* Sesleria albicans *(open circles). For each species the photon flux at the leaf surface was ca. 2000 μmol m^{-2} s^{-1} (From Long, 1976).*

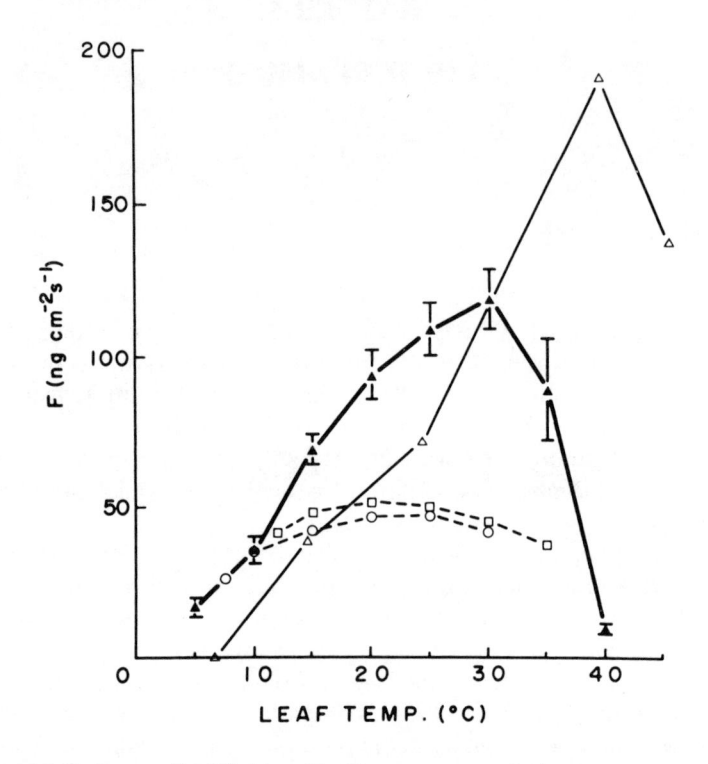

Berry and Björkman (1980) identify three groups of plants:
(i) species adapted to low temperature which cannot acclimate to high temperature
(ii) species adapted to high temperature which cannot acclimate to low temperature
(iii) species, including several evergreen shrubs, which photosynthesize throughout the year and acclimate over a wide range of temperatures.

All the above groups of plants have some capacity to acclimate to changing ambient temperature as their optimum temperature for photosynthesis will change by 1–3°C for every 5°C change in growth temperature (Fig. 5.2; Pisek et al., 1969; Mooney, 1978). The substantial changes in optimum temperature which occur within a few days following the transfer of plants from one temperature extreme to another in controlled environments suggest that many species have considerable potential for acclimation to changing growth temperatures in the field where changes of temperature are less abrupt (Rook, 1969; Mooney and Harrison, 1970; Sawada and Miyachi, 1974; Hickelton and Oechel, 1976). Slatyer and Morrow (1977) have observed a close correlation between the optimum temperature for photosynthesis in *Eucalyptus* and the mean maximum temperature of the ten days prior to the date of measurement. Such adaptations of the temperature response curve to prevailing environmental conditions could also have a role in plant survival as the raising or lowering of the high and low temperature compensation points, respectively, delays the onset of irreversible effects of temperature stress and a consequent loss of productivity. Besides phenotypic plasticity, ecotypic differences occur with respect to optimum temperature (Billings *et al.*, 1971; Slatyer and Ferrar, 1977).

Fig. 5.2 The effect of growth temperature on the rate and temperature dependence of light-saturated net CO_2 uptake for a number of C_3 and C_4 species illustrating temperature acclimation through a shift in optimum temperature for photosynthesis. The "hot" growth regimes were 40°C for Atriplex glabriuscula *(C_3), A.* sabulosa *(C_4), and A.* hymenelytra *(C_4), 43°C for A.* lentiformis *(C_4) and 45°C for* Tidestromia oblongifolia *(C_4) and* Larrea divaricata *(C_3). The "cool" growth regimes were: 23°C for A.* lentiformis, *20°C for L.* divaricata, *and 16°C for the other species (reproduced with permission from Björkman et al., 1980).*

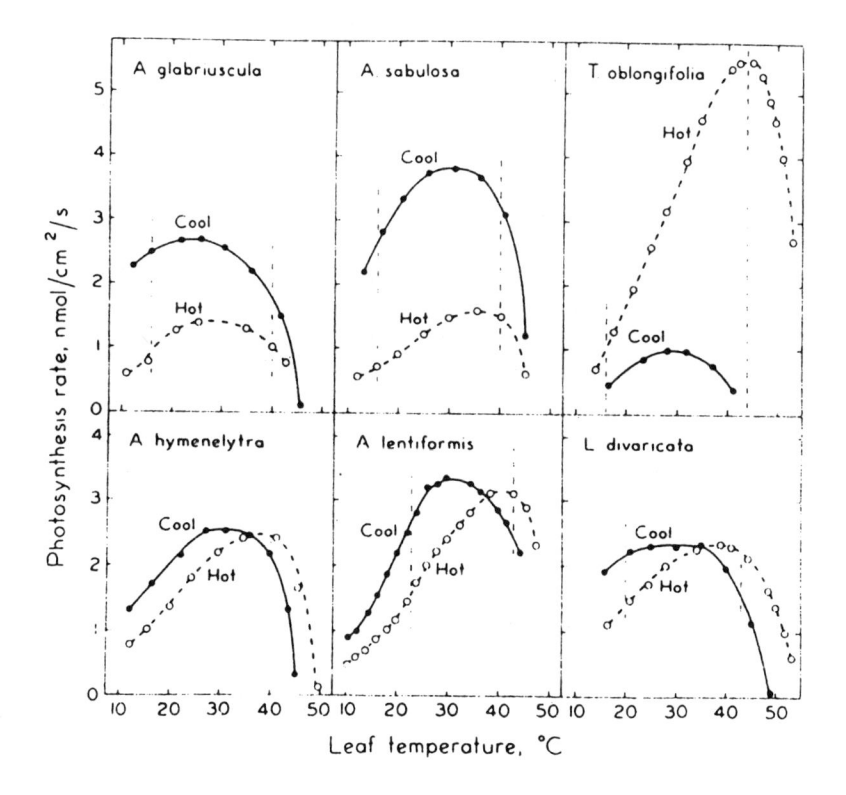

5.2 REVERSIBLE EFFECTS

The underlying mechanisms which determine the shape of the temperature response curve are a complex integration of biophysical and biochemical processes all of which are influenced by temperature (Treharne and Nelson, 1975; Berry and Björkman, 1980). A major biophysical limitation on photosynthesis is stomatal conductance but it is now clear from a number of studies that stomatal closure is not the primary cause of the decline in photosynthesis at supraoptimal temperatures (Bauer, 1979; Raschke, 1975; Björkman et al., 1980). An important factor in the biochemistry of C_3 plants is the increase in the ratio of photorespiration : photosynthesis with increase in temperature which results from a more rapid decrease in the affinity of Ru*bis*CO for CO_2 than O_2 (Laing et al., 1974; Badger and Collatz, 1977). The higher optimum temperature of C_4 species, all of which suppress photorespiration, is consistent with this hypothesis and the measured response of C_4 plants. The precise explanation for the differential decrease in the affinity of Ru*bis*CO for its substrates is not clear, but may relate to differential effects of temperature on the solubities of O_2 and CO_2.

The sharp decline in photosynthesis observed at supraoptimal temperatures in C_3 plants cannot be fully explained by the increase in photorespiration (Berry and Björkman, 1980). Current thinking suggests that for both C_3 and C_4 plants it is related to a decline in the rate of supply of Ru*b*P with increasing temperature linked to a marked decline of coupled photosynthetic electron transport and a reduced supply of NADPH (Nolan and Smillie, 1976; Armond et al., 1978; Björkman et al., 1978; Farquhar, 1979; Berry and Björkman, 1980).

In contrast the decline in photosynthetic performance at suboptimal temperatures is more a function of the activity of rate limiting dark reactions in both C_3 and C_4 plants (Björkman, 1973). In C_3 plants photosynthetic rate is best correlated with the activity of fructose 1,6-*bis*phosphatase activity, a key step of the reductive pentose phosphate cycle, and in C_4 plants with Ru*bis*CO (Björkman and Badger, 1977; Pearcy, 1977, Portis et al., 1977; Björkman and Badger, 1979; Björkman et al., 1980). Berry and Björkman (1980) considered these differences to be entirely consistent with the metabolic differences between C_3 and C_4 plants as Ru*bis*CO is much less temperature dependent at ambient CO_2 concentrations than at the rate saturating concentrations which prevail in the bundle sheath chloroplasts of C_4 plants.

5.3 IRREVERSIBLE DAMAGE

The reversible effects of temperature on photosynthesis are the inevitable expression of the complex relationships between photosynthetic and photorespiratory processes and the temperature-related chemical properties of enzymes. As temperature is not an environmental factor which can be controlled in the field, growth and biomass production will be determined by diurnal and seasonal changes of temperature in concert with other limiting factors. A further constraint on production is the irreversible damage or injury to the photosynthetic system caused by temperature extremes which then prevent expression of potential productivity even on return to optimum growth conditions (Fig. 5.3).

Plants can be divided into chilling-sensitive species, which can show signs of cell degradation well above freezing point and include most C_4 plants, and chilling-tolerant species which remain photosynthetically active as long as the cells are not frozen (Larcher, 1981). Distinct differences in the responses of plants at high temperature are also observed, since some species are able to maintain cell integrity up to much higher temperatures than others.

Analysis of temperature lesions has been more frequently studied *in vitro* and it is possible that partial photosynthetic processes are less temperature sensitive *in vivo* in the intact system. It is also notoriously difficult to separate the effects of temperature and drought stress. Results from such experiments need cautious interpretation.

The tolerance of plant membranes to heat damage exceeds that of photosynthesis and the tolerance of the chloroplast envelope is greater than that of the photosynthetic membrane (Berry *et al.*, 1975; Krause and Santarius, 1975; Björkman *et al.*, 1980). The reasons for the irreversible decline in light saturated photosynthesis under heat stress may therefore be photosynthetic in origin and related to the lipid properties of the membranes which support photosynthetic electron transport (Berry and Raison, 1981; Öquist, 1983). For example, photosystem II showed a temperature sensitivity similar to that of whole leaf photosynthesis and was probably associated with a failure of the water-splitting apparatus (Björkman *et al.*, 1978; Bauer and Senser, 1979; Berry and Björkman, 1980). A concurrent decline in photosynthesis at high temperature is also correlated with a considerable reduction in quantum yield (Björkman, 1975; Björkman *et al.*, 1976; Pearcy *et al.*, 1977; Schreiber and Berry, 1977; Ludlow, 1980). This was considered a result of the breakdown of energy transfer between the light harvesting molecules, the reaction centre, and the electron transport system (Armond *et al.*, 1978). Since the strength of hydrophobic bonds increase and the strength of hydrophilic bonds decrease at high temperature, the distance between the light harvesting molecules and the reaction centre increases and disrupts chloroplast function (Berry and Björkman, 1980; Raison *et al.*, 1980).

The stability of most enzymes of the dark reactions, including Ru*bis*CO, exceed that of photosynthesis at high temperature (Tieszen and Sigurdson, 1973; Yordanov and Vasileva, 1976; Björkman and Badger, 1977; Björkman *et al.*, 1978) including Ru*bis*CO from spinach and a number of grasses. The only enzymes which express stabilities similar to that of photosynthesis are the light activated enzymes NADP glyceraldehyde 3-P dehydrogenase, ribulose 5-P kinase and NADP malate dehydrogenase (Anderson, 1975; Hatch, 1977; Björkman and Badger, 1979, Björkman *et al.*, 1980). This may result from heat inhibition of photosystem II

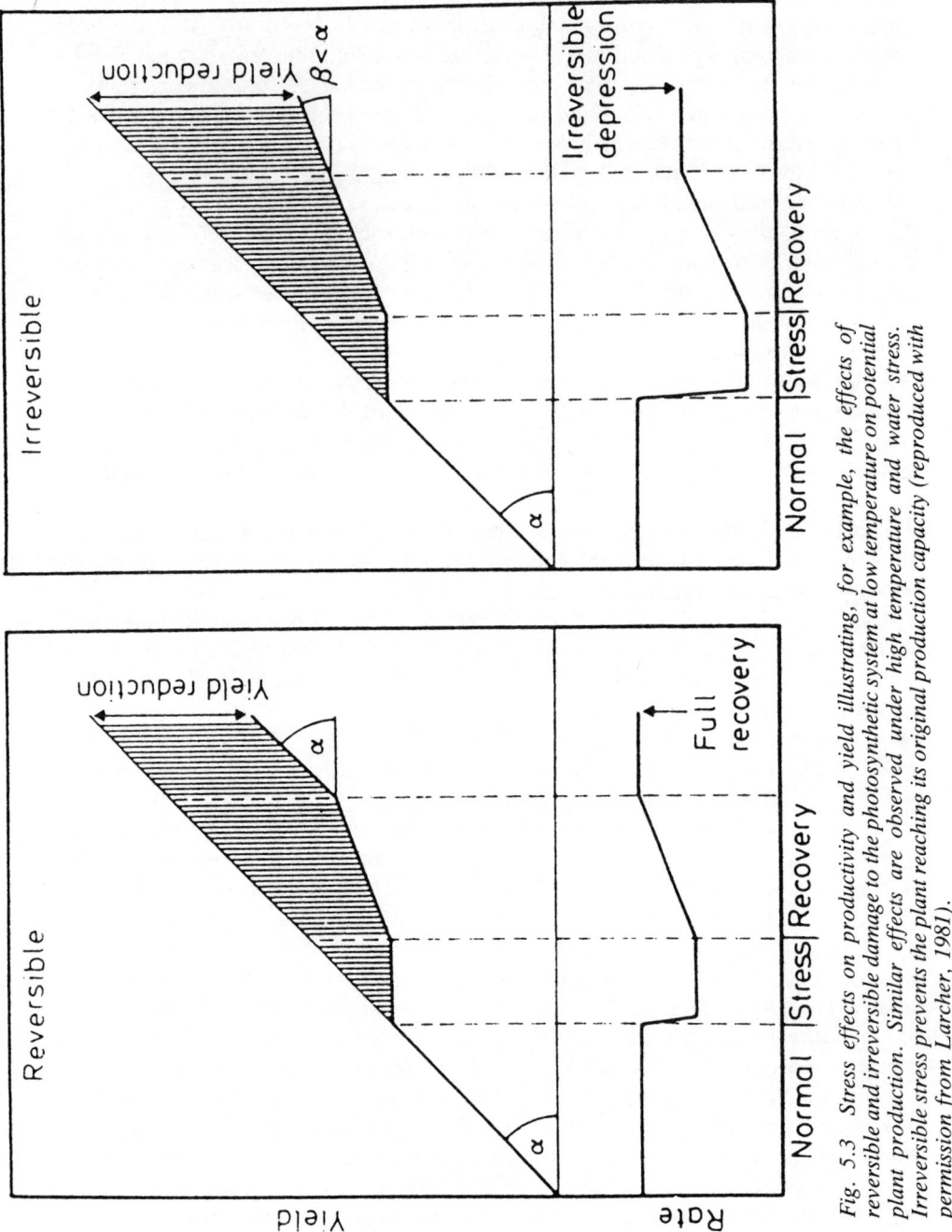

Fig. 5.3 Stress effects on productivity and yield illustrating, for example, the effects of reversible and irreversible damage to the photosynthetic system at low temperature on potential plant production. Similar effects are observed under high temperature and water stress. Irreversible stress prevents the plant reaching its original production capacity (reproduced with permission from Larcher, 1981).

(Berry and Björkman, 1980). However, there is no basis for presuming that the stability of these enzymes is a function of their activity.

The deleterious effects of low temperature differ from those of high temperature and there is no evidence to suggest that characteristics of the photosynthetic system can in any way determine the capacity of plant tissue to survive freezing. Nevertheless, hysteretic or irreversible effects on photosynthesis occur at temperatures well above the freezing point and these are not causally related to the decline in water status at low temperatures (Taylor and Rowley, 1971; Bagnall, 1979). Furthermore a combination of low temperature and high light can cause inhibition of chlorophyll synthesis, photobleaching of chlorophyll and photoinhibition of photosynthesis (Fig. 5.3; McWilliam and Naylor, 1967; Taylor and Rowley, 1971; Van Hasselt, 1972; Slack et al., 1974; Van Hasselt and Strikverde, 1976; Powles et al., 1980; Van Hasselt and van Bierlo, 1980; Baker et al., 1983; Long et al., 1983; Powles et al., 1983).

The after effects of chilling include a reduced stomatal conductance (Tschäpe, 1972; Ivory and Whiteman, 1978; Long et al., 1983). In one instance this was related to an increased sensitization of the guard cells to the internal CO_2 concentration (Drake and Raschke, 1974). In the chloroplast much of the photosynthetic process occurs on or is closely associated with the membranes of the thylakoids and stromal lamellae. These are composed of a phospholipid matrix which must remain fluid for the proper functioning of photosynthetic electron transport. At low temperature the fluidity of the membrane declines and lateral phase separation of the gelled from the remaining liquid components occurs (Linden et al., 1973; Wolfe, 1978; Lyons and Breidenbach, 1979; Raison, 1980). These changes are clearly complex, but to some extent correlated with the fatty acid composition of the membrane, and in particular their degree of unsaturation (Steponkus, 1981; Öquist, 1983). Chilling sensitive plants have a lower ratio of unsaturated : saturated fatty acids (Tajima, 1971). Phase separation together with abrupt changes in the activity of chloroplasts and mitochondria occur at higher temperatures in these plants (Pike and Berry, 1979; Pike et al., 1979; Raison et al., 1979). Acclimation to chilling temperature is associated with an increase in the ratio of unsaturated : saturated fatty acids (Wilson, 1979).

It is not surprising therefore that a decline in the activities of electron transport and the enzymes of carbon metabolism are observed at low temperature (Berry and Raison, 1981). For example, the capacities for electron transport *in vitro* through PS I, PS II and PS I plus PS II were all inhibited in Scots pine, and a more severe destruction of chlorophyll occurred in the photosystems than in the light harvesting chlorophyll a/b complex (Martin et al., 1978; Öquist, 1981). Analysis of chlorophyll fluorescence induction suggests that similar changes in electron transport occur *in vivo* in maize subjected to chilling temperatures (Baker et al., 1983). Ru*bis*CO activity was low in wheat at low temperature though substantial activities were present in Scots pine even under severe winter stress (Sawada et al., 1974; Gezelius and Hallén, 1980). This accords with evidence that Ru*bis*CO may be synthesized in more stable forms at low temperature (Huner and MacDowell, 1979). Pyruvate orthophosphate dikinase, the enzyme regenerating the primary CO_2 acceptor in C_4

plants showed a marked increase in its activation energy at 10°C in maize suggesting a change in tertiary structure at this temperature (Taylor *et al.*, 1974; Shirahasi *et al.*, 1978). Differences in the cold lability of this enzyme from maize cultivars in Japan were positively correlated to their northern limits (Sugiyama and Boku, 1978). The rate of low temperature inactivation also differed according to the species from which the dikinase was extracted (Sugiyama *et al.*, 1979). Caldwell *et al.* (1977) suggested that C_4 plants which decarboxylate through NADP-linked malic enzyme were more temperature sensitive than those plants which utilize NAD-linked malic enzyme, but this has not been further substantiated. The effects of low temperature on C_4 photosynthesis have been reviewed by Long (1983).

CHAPTER 6
Soil Factors

6.1 WATER

6.1.1. Importance of measurement

Water is the most abundant component of plant cells contributing 60–90% or even more of the total fresh weight. It is fundamental to photosynthesis as a reactant, as a milieu for biochemical reactions and has a passive role in transpiration. The availability and utilization of water are major factors influencing the growth and yield of crops and forest stands even in temperate or humid climates (Tazaki *et al.*, 1980). Periods of drought are well known for their devastating repercussions on production in arid and semi-arid zones.

The rate of net photosynthesis declines under water stress and may cease completely should severe water deficits develop. The cessation of leaf area expansion because of low turgor pressures (Boyer, 1970a), the mobilization of carbohydrate reserves to offset the loss of new photosynthate (Fischer, 1973), ABA production which inhibits phloem loading and the diversion of photosynthate or osmoregulation (Mansfield and Wilson, 1981) are also features of plant water deficits which decrease productivity. To some extent the effects of water stress on productivity are compensated for by the mobilization of storage compounds into harvestable products. It is almost certain however that the effects of drought on current photosynthesis contribute in part to the loss of crop yield (Boyer, 1976a,b).

The quantity or activity of water available for plant growth can be expressed with respect to the soil, but for the purposes of this chapter it is more convenient to consider the water status of the plant itself. This may be expressed in terms of the relative water content (RWC) where:

$$RWC = \frac{\text{fresh weight} - \text{dry weight}}{\text{saturated fresh weight} - \text{dry weight}}$$

or preferably in terms of the total water potential or Gibbs free energy of water in the plant tissue relative to that of pure water.

The total water potential, Ψ_{plant} consists of two major components, the osmotic potential (Ψ_π) arising from the presence of dissolved solutes in the cell, and the turgor or wall potential (Ψ_p) arising from the pressure exerted on the cells by their walls. Each is expressed in pressure units (MPa). Thus:

$$\Psi_{plant} = \Psi_\pi + \Psi_p$$

Since water potential depends simply on the Gibbs free energy content of water, it may be calculated for any water containing system, including the soil and the atmosphere. Thus, a common method of describing water status is provided for both the plant and its environment. Water potential is a useful index and may be directly related to the molecular activity of water. However, this does not imply that the reduction in activity of water during stress has a direct effect *per se* on photosynthesis. Indeed, neither Ψ_π nor Ψ_p were correlated with the fall in photochemical activity of chloroplasts isolated from *Helianthus annuus* plants at low water potential (Boyer and Potter, 1973; Potter and Boyer, 1973).

6.1.2 Water stress and photosynthesis

To obtain CO_2 for photosynthesis, leaves expose wet surfaces *viz.* the cell walls of the substomatal cavity to the atmosphere, and suffer evaporative water loss as a consequence. It is often considered that this flux of water is not essential to the plant, but is an inevitable prerequisite for obtaining a major substrate i.e. CO_2 for photosynthesis (Berry, 1975). Evaporative cooling, nevertheless, often accounts for a considerable proportion of heat dissipation by vegetation (Uchijima, 1976). This is probably essential for maintaining equable temperatures for photosynthesis particularly under water stress. Therefore it is perhaps more appropriate to describe the water lost in exchange for CO_2 as being used for evaporative cooling (Good and Bell, 1980). Since water and CO_2 follow the same diffusion pathways transpiration is beneficial to photosynthesis at two levels.

Each major resistance in the diffusion pathway of CO_2 from the atmosphere to the sites of carboxylation within the mesophyll may increase with water stress. The curling of leaves increases the boundary layer resistance. Loss of turgor pressure in the guard cells leads to closure of the stomata whilst increased ABA levels under water stress inhibit stomatal opening, an effect which may persist several days beyond the return of the plant to a higher water potential. In both instances stomatal resistance increases. Water stress also increases mesophyll resistance. Measured changes of these resistances during the decline of photosynthesis under water stress are used to distinguish their relative importance.

The fall in photosynthesis at low water potential is accompanied by antiparallel changes in stomatal resistance in many species (see Boyer, 1976a,b for a comprehensive list of examples). In some instances the correspondence between photosynthesis and transpiration (inversely proportional to resistance at constant vapour pressure deficit) has been almost perfect and seemingly decisive evidence for exclusive stomatal control of photosynthesis during water stress (Brix, 1962;

Boyer, 1970b; Beadle *et al.*, 1973). As Boyer (1976a,b) has comprehensively argued however, this conclusion would be invalid if photosynthesis had not been limited by diffusion through the stomata before the stress was imposed.

There is much evidence that partial photosynthetic processes of non-stomatal origin are inhibited at low water potentials (Kriedemann and Downton, 1981). Boyer (1971) found that stomatal closure was not sufficient to account for the fall in net photosynthetic rate in *Helianthus annuus* at low water potential. A parallel fall in the rate of photosynthetic electron transport was observed in the same experiment though the eventual decline of photosynthesis to zero was later found only to be correlated with falls in cyclic and non-cyclic photosphorylation (Fig. 6.1, Keck and Boyer, 1974). This is possibly effected as a result of the concentration of Mg^{2+} to inhibitory levels within the chloroplasts (Boyer and Youmis, 1983). The quantum yield was also more than halved as water potential fell from -0.4 to -1.5 MPa in *Helianthus annuus* and from -1.0 to -3.6 MPa in the desert shrub *Larrea divaricata* (Mohanty and Boyer, 1976; Mooney *et al.*, 1977). Reductions in photosystem activity were observed in other species over a range of dessication treatments in *Triticum aestivum, Beta vulgaris* and *Gossypium hirsutum* (Todd and Basler, 1965; Nir and Poljakoff-Mayber, 1967; Fry, 1970) In *Picea sitchensis*, the activities of PS I and PS II were wholly independent of water potential, even in needles which were severely desiccated (Beadle and Jarvis, 1977). High residual activities of both photosystems were observed in *Helianthus annuus*, even in air-dried tisue (Keck and Boyer, 1974). The ultrastructural appearance of chloroplasts at low water potentials confirms the ability of photosynthetic membranes to retain their structural integrity even under severe dessication (Giles *et al.*, 1974; Fellows and Boyer, 1978). However, the mesophyll chloroplasts of C_4 plants are more readily disorganized at low water potential than are the chloroplasts in the bundle sheath (Giles *et al.*, 1974). Loss of chlorophyll from the light harvesting chlorophyll a/b-protein complex was observed in *Zea mays*, a C_4 species, under water stress (Alberte *et al.*, 1977).

Activities of enzymes of the dark reactions decline at low water potential but not to the extent that photosynthetic rates *in vivo* would be limited. For example, there was no reduction in the activity of Ru*bis*CO isolated from leaves as the water potential of the leaf (Ψ_{leaf}) decreased to -2.6 MPa in *Picea sitchensis* and only small reductions in Ru*bis*CO activity were observed in similar experiments with *Gossypium* spp., *Pisum sativum* and *Phaseolus vulgaris* seedlings, and *Triticum aestivum* and *Hordeum sativum* (Jones, 1973; Johnson *et al.*, 1974; Lee *et al.*, 1974; Beadle and Jarvis, 1977; O'Toole *et al.*, 1977). Changes in PEP carboxylase and ribulose 5-phosphate kinase were also insufficient to explain observed reductions in photosynthesis in *Hordeum sativum* and *Sorghum bicolor* (Huffaker *et al.*, 1970; Shearman *et al.*, 1972). The distribution of assimilated ^{14}C between the products of photosynthesis alters under water stress (Fig. 6.2). In water stressed *Zea mays* (C_4) there was a greater proportion of ^{14}C in organic acids compared to well-watered controls. In *Helianthus annuus* (C_3), water stress resulted in an increased proportion of ^{14}C in glycine and serine, and a decreased proportion in sugars (Lawlor and Fock, 1977; Lawlor, 1979). These results were consistent with a

Fig. 6.1 Activity of Helianthus annuus *chloroplasts from leaves that had been desiccated to varying degrees showing the falls in the rates of photosynthetic electron transport and photophosphorylation at low leaf water potential (1 bar = 0.1 MPa). (a) photosystem I electron transport; (b) electron transport from water to methyl viologen i.e. photosystem I plus photosystem II; (c) cyclic electron flow and (d) non-cyclic electron flow. (1 bar = 0.2 MPa, reproduced with permission from Keck & Boyer, 1974).*

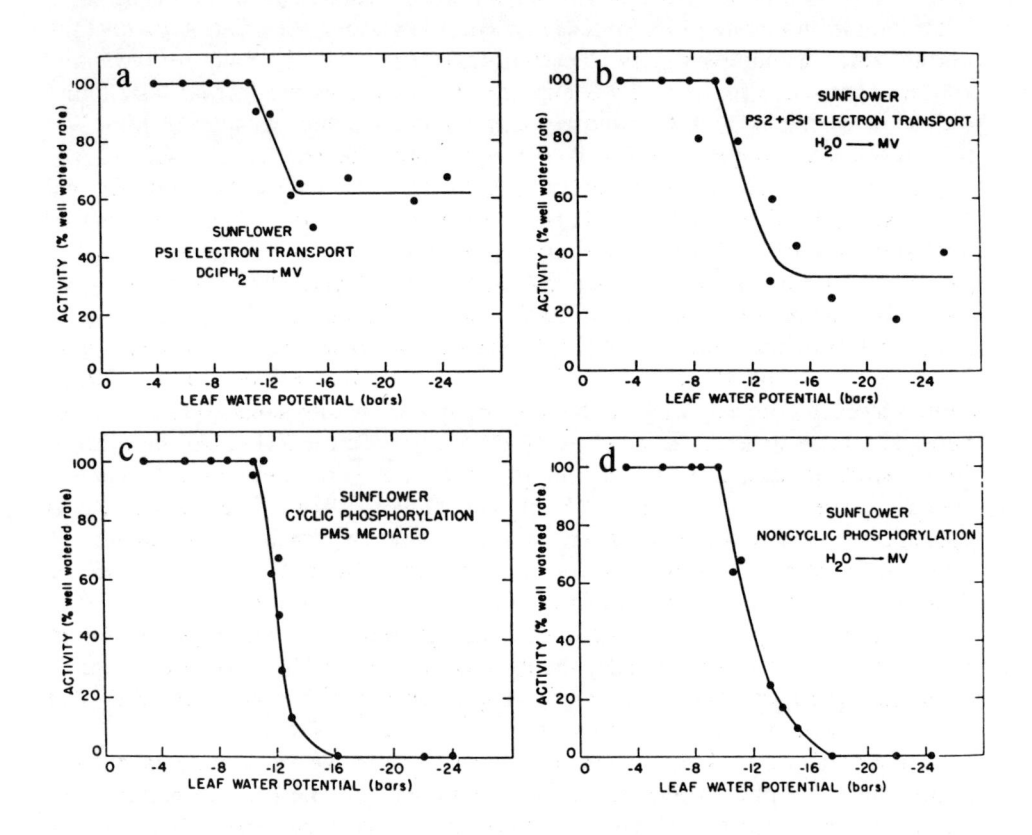

decrease in photosynthetic rate in the C_4 species and a relative increase in the ratio between photorespiration and photosynthesis in the C_3 species. Proportionate increases in photorespiration may be a function of an increase in the ratio of oxygenation : carboxylation of RubP with stress which is caused by a fall in the intercellular CO_2 concentration following stomatal closure (Laing *et al.*, 1974; Lawlor, 1979). Increase in the CO_2 compensation point at low water potential in both C_3 and C_4 plants suggests that photorespiration increases with stress (Glinka and Katchansky, 1970; Lawlor, 1976a).

Many, though not all, of the above experiments have been done by subjecting plants to treatments which induce a low Ψ_{leaf} within periods extending from a few hours to a few days. In the field, changes in Ψ_{leaf} are usually much slower. Observed effects of rapidly applied stress treatments are remarkably similar to the

SOIL FACTORS

photoinhibition of photosynthesis at zero CO_2 concentration and 1% O_2 (Fig. 6.3), suggesting that rapid stress treatments may in some way interfere with the utilization of photochemical energy even in the presence of CO_2 (Osmond et al.,

Fig. 6.2 The distribution of assimilated ^{14}C in soluble compounds from Helianthus annuus *(C_3) and* Zea mays *(C_4) leaves at different water potentials (10^5 Pa = 0.1 MPa) showing the changes which occur under water stress (reproduced with permission from Lawlor, 1979).*

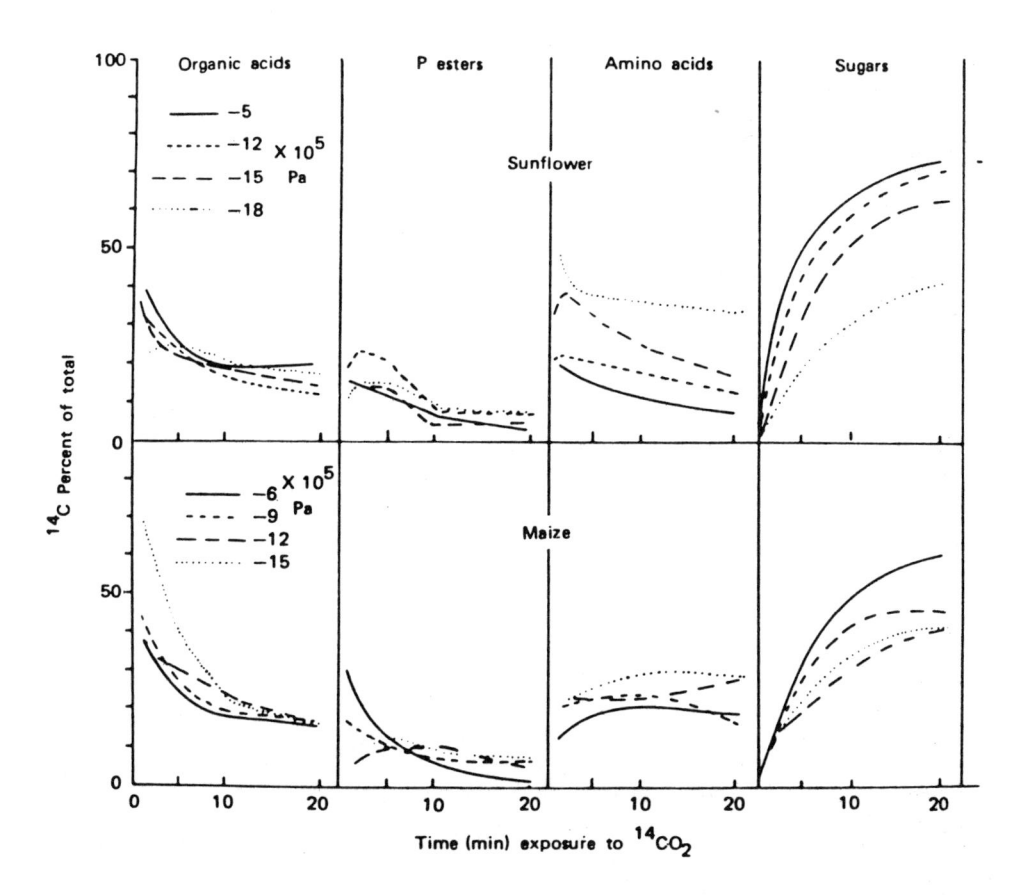

1980). Slowly applied stress treatments which are similar to those found under field conditions may lead to more co-ordinated adjustments at the stomatal and mesophyll level. Further, it is necessary to exercise care when interpreting the effects of water stress on photosynthesis *in vivo* from experiments *in vitro*. If low water potentials are simulated by stressing chloroplasts osmotically *in vitro*, the added solutes, which are not perfectly inert, may produce effects beyond those of lowered water potential. The activity of chloroplasts previously stressed *in vivo* may, when measured *in vitro*, solely reflect their level of hydration (Darbyshire and Steer, 1973; Beadle and Jarvis, 1977).

Fig. 6.3 The relationship between CO_2 fixation and intercellular CO_2 concentration (a, c) or incident quantum flux density (b, d) of Phaseolus vulgaris *(C_3) at 30°C and in normal air (330 cm^3 m^{-3} CO_2, $21 dm^3$ m^{-3} O_2). In the upper and lower figures the observations were made before, immediately after and 24 h after exposure to a quantum flux density of 2000 umol m^{-2} s^{-1} for 3 h at 30°C in an atmosphere containing either $21 dm^3$ m^{-3} oxygen with $70 cm^3$ m^{-3} CO_2 or $1 dm^3$ m^{-3} oxygen, with zero CO_2, respectively. These effects of photoinhibition are similar to those observed under water-stress treatments (reproduced with permission from Osmond et al., 1980).*

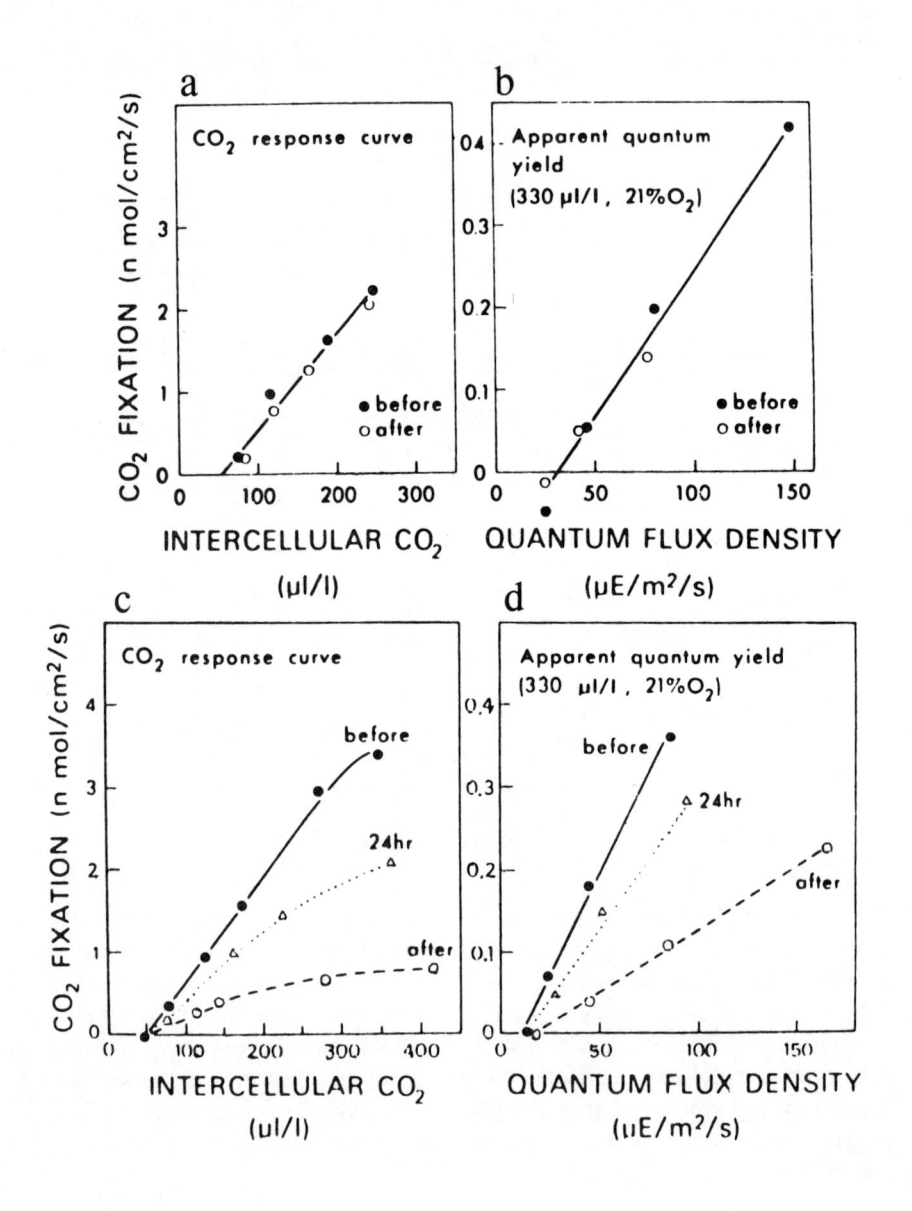

From the current state of knowledge it can be concluded that photosynthesis declines at low water potential from the simultaneous effects of both stomatal closure and a decreased chloroplast activity. Gross changes in photosynthesis may be distinguished in terms of limitations at stomatal and mesophyll level (Fig. 6.4, Beadle et al., 1981; Jones and Fanjul, 1983). The time period over which r_s and r_m decrease following water stress and their relative importance vary both within and between species. There may be considerable adjustments of both r_s and r_m to avoid large decreases in the intercellular CO_2 concentration (C_i) which would otherwise occur at low Ψ_{leaf} (Bradford and Hsaio, 1982). Low C_i causes photoinhibition of photosynthesis under the environmental conditions which promote water stress, *viz.* high light and temperature extremes. It has been proposed that the recycling of CO_2 by photorespiration in C_3 plants, metabolite transfer in C_4 plants, and recycling of CO_2 fixed in the dark and photorespiration in CAM plants, serve to dissipate the excess photochemical energy present while these stress conditions persist (Osmond et al., 1980).

The effort expended in measuring both the effects of water stress on the development of the photosynthetic apparatus and the recovery of photosynthesis from water deficits has been negligible compared to that expended in measuring the immediate effects of water stress. Frequent rapid recovery of photosynthesis after a period of stress confirms that chloroplast function is maintained to the extent that the rate of photosynthesis may even exceed that of leaves of similar chronological age which were not subjected to a stress treatment (Ludlow and Ng, 1974). Ludlow (1976) hypothesized that leaf ageing was suspended during periods of stress and that on watering plants still had the capacity for photosynthetic rates commensurate with their physiological age. Recovery of photosynthesis is often delayed owing to after effects of stress which prevent full stomatal opening (Sanchez-Diaz and Kramer, 1971; Loveys and Kriedemann, 1973; Kriedemann and Loveys, 1974). The period of this delay is a function of the degree and duration of the stress treatment (Fischer et al., 1970).

6.1.3. Water stress and bioproductivity

The precise contribution to the loss of biomass production through the reduction of photosynthesis during periods of water stress is difficult to quantify. Stress periods differ in length and intensity both in different environments and in different seasons, and may coincide with different growth stages which vary in their sensitivity to water stress. There is a good correlation between the reductions in transpiration and biomass production resulting from drought (see Schulze and Hall, 1981). Since there is much evidence to suggest that biomass production is correlated with water use, particularly in determinate crops, even mild water stress would cause reduction in productivity (Hanks et al., 1969; Shouse et al., 1977; Turk and Hall, 1980a,b), but by how much?

PHOTOSYNTHESIS IN RELATION TO PLANT PRODUCTION

$g_s = 1.0 - 1.0e^{-2.5(\psi-(-2.62))}$
$F_c = 1.0 - 1.0e^{-3.1(\psi-(-2.64))}$
$g_m = 1.0 - 1.0e^{-3.7(\psi-(-2.65))}$

Fig. 6.4 Relative stomatal conductance (— — —), relative mesophyll conductance (······) and predicted net photosynthesis at constant mesophyll conductance (———) as a function of xylem water potential in Sitka spruce. F_c represents the reduction in $\hat{F}_c$ caused by stomatal closure and the hatched area the reduction in F_c as a result of the reduction in mesophyll conductance or chloroplast activity. Leaf temperature was 20°C; C_a was 300 cm^3 m^{-3}. (Reproduced with permission, Beadle et al., 1981)

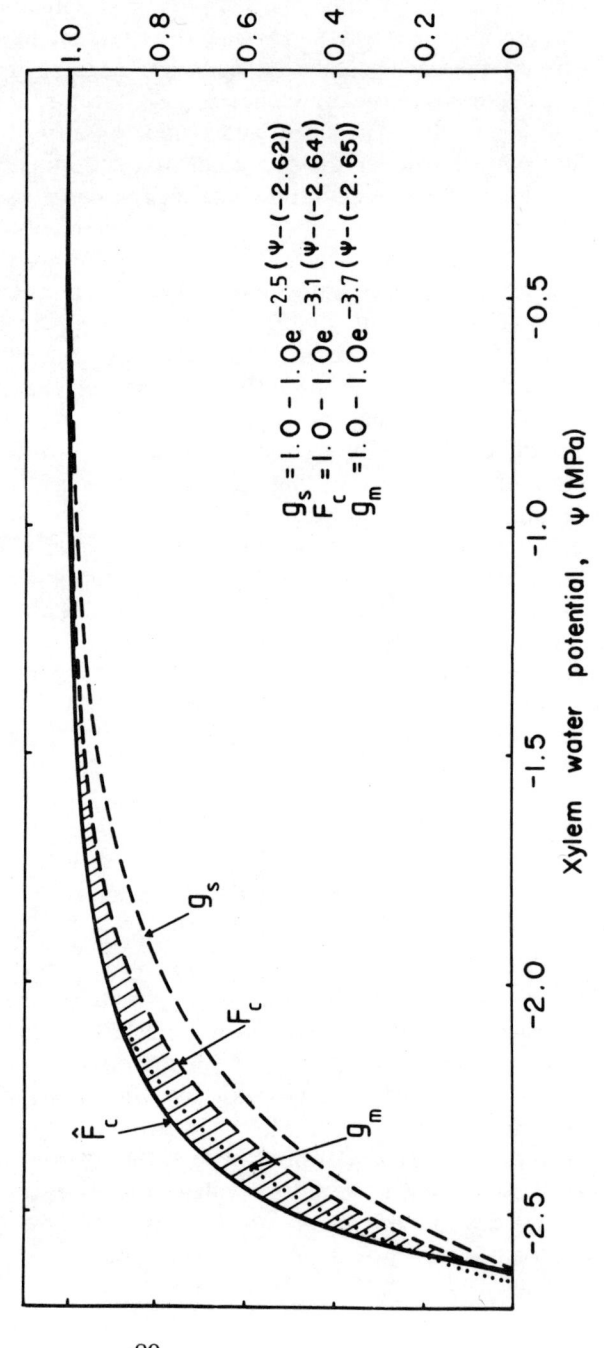

The accumulation of biomass can be summarized in terms of three processes (Day, 1981)

$$\begin{array}{c}\text{total biomass}\\ \text{production}\end{array} = \begin{array}{c}\text{light}\\ \text{intercepted}\end{array} \times \begin{array}{c}\text{efficiency of}\\ \text{photosynthesis}\end{array} \times \begin{array}{c}\text{fraction remaining}\\ \text{after respiration}\end{array}$$

In a comparison of unirrigated and irrigated *Hordeum sativum* the major factor which decreased biomass production in the unirrigated crop was a 40% reduction in the light intercepted which resulted from a decrease in green leaf area and a shortened growing season (Day, 1981). In the same experiment, the estimated effect of stomatal closure on the unirrigated plants was a rate of photosynthesis 7% lower than in the irrigated plants (Legg *et al.*, 1979). In many plants the economic yield is only a part of the total biomass production, e.g. cereals and root crops. In these plants, the proportionate decrease in economic yield resulting from water stress is usually less than that in total biomass production. For example, McPherson and Boyer (1977) subjected *Zea mays* plants to a continuous period of water stress between tasseling and harvest. Although photosynthetic rate was approximately zero throughout this period, grain yield was 47–67% of controls. Translocation can clearly be less affected by water stress than photosynthesis (Sung and Krieg, 1979). In cereals subjected to water stress after anthesis there is some doubt as to whether the total contribution to grain filling from reserves already present in the crop at anthesis increases (Gallagher *et al.*, 1975; McPherson and Boyer, 1977) or remains constant, i.e. a proportional change only (Bidinger *et al.*, 1977). Fischer (1980) maintains that higher estimates result from a failure to account for weight losses due to damage and disease during post-anthesis growth. The reduction in current photosynthate supply is, at least in part, counteracted by the mobilization of storage compounds and emphasizes the importance of the integrated photosynthetic accumulation as a determinant of yield. Adaptation of photosynthesis to plant water stress varies between and within species. The effects of water deficit on photosynthesis occurred later during a drought period in one variety of *Glycine max* compared to a second, though this difference was probably not related to differences in photosynthetic efficiency at chloroplast level under stress (Turner *et al.*, 1978).

The water lost by transpiration and biomass production in terms of dry matter or CO_2 fixed are linked through the term water use efficiency (e.g. dry weight gained per unit mass of water transpired). There is no widely accepted definition of water use efficiency (WUE) nor is it possible to assign unique values to any one species. Assuming an infinitely high affinity for CO_2, the maximum WUE is 30 mg CO_2/g H_2O at 25°C and 50% R.H. (relative humidity) (Fischer and Turner, 1978), but in practice this figure is much lower indicating a decreased WUE. In general, C_4 plants have much higher WUE's than C_3 plants (Schantz and Piemiesel, 1927; Good and Bell, 1980) consistent with their higher individual leaf photosynthetic rates while CAM plants have the highest WUE's. As biomass production is proportional to water use, it would appear that WUE is a constant for a species in a given environment (Day, 1981). Maximum yields will only be realized therefore by supplying sufficient water to meet evaporative demand during the growing season or by real increases in WUE through higher rates of crop photosynthesis per unit of water transpired. The water use efficiencies of some natural ecosystems are considered by Webb *et al.* (1978).

6.2 SALINITY

6.2.1 Salinity in the environment

Unlike light, temperature and water, the level of soil salinity is not a feature influencing productivity in all terrestrial environments. Perhaps as a result, the specific effects of salinity on photosynthesis have received less attention. The deleterious symptoms observed in plants at critical salt concentrations are also the result of a complex series of interactions and there is no *prima facie* case for expecting a single lesion (Wyn Jones, 1981) or that photosynthesis is necessarily a primary site of action.

"Saline soil" is normally used in plant physiology to indicate a soil with an electrolyte concentration which is inhibitory to the growth of crop plants. Typically these are soils dominated by NaCl or Na_2SO_4. Although of lesser importance, other elecrolytes have been shown to be present at inhibitory concentrations in certain soils, in particular $MgSO_4$, $CaSO_4$, $MgCl_2$, KCl and Na_2CO_3 (Flowers et al., 1977; Szabolc, 1979). "Saline soils" have been well maped in Australia and Europe and occupy ca. 3×10^6 km^2 (Northcote and Shene, 1972). Minimum estimates suggest a further 10^6 km^2 occur in the rest of the World, excluding the hot deserts (Northcote and Shene, 1972; Murdie, 1974; Szabolc, 1974, 1979; Croughan and Rains, 1982). In addition to natural saline soils, secondary salinization is continually adding to this total through irrigation and drainage practices. Two-thirds of the world's canal-irrigated land (0.15×10^6 km^2) is becoming saline resulting in a large and growing rate of loss of agricultural land (Mohammed, 1978). In the Punjab, irrigation had by 1960 caused the salinization of some 25% of the 51 000 km^2 of agricultural land. Similar losses of agricultural land by irrigation with poor quality water have been reported for Maharashtra, India (Boyko, 1966; Joshi, 1976). At a conservative estimate 400 km^2 of formerly productive agricultural land are lost annually by secondary salinization (Boyko, 1966).

6.2.2. Stress factors and tolerance

Wyn Jones (1981) has summarized three major factors which potentially limit the growth of plants in saline habitats: (1) Water stress due to the osmotic effects; (2) Specific ion toxicity; (3) Ion imbalance stress or induced nutrient deficiency.

The potential productivity in a given saline environment however is affected to different degrees according to the plant species and other environmental and plant factors (Maas and Hoffman, 1977). Halophytes represent the native flora of saline lands (Jennings, 1968) and generally, plants classified as halophytes will survive salinities in excess of 500 mM. Most crop plants are glycophytes (non-halophytes, Maas and Hoffman, 1977) which will not survive at this concentration of salts, and show some reduction in growth as a result of any increase in NaCl concentration above 20 mM (Fig. 6.5). For example, the maximum concentration at which no growth can occur and above which the plant dies is as low as 100 mM NaCl for many fruit crops. Other crops are more tolerant and selected cultivars of *Hordeum sativum* have been shown to survive salinities up to 300 mM (Epstein and Norlyn, 1977; Greenway and Munns, 1980).

Fig. 6.5 Growth responses of different species to salinity after 1–6 months at high Cl⁻ in the external medium. Curve 1: Sueda maritima*; Curve 2: sugar beet (0–150 mM Cl^-_{ext}) and* Spartina townsendii *(150–700 nm $NaCl_{ext}$); Curve 3: cotton; Curve 4: beans. Group I are halophytes; Group II are halophytes and non-halophytes with differing sensitivity to salinity; Group III are salt sensitive non-halophytes (reproduced with permission from Greenway and Munns, 1980).*

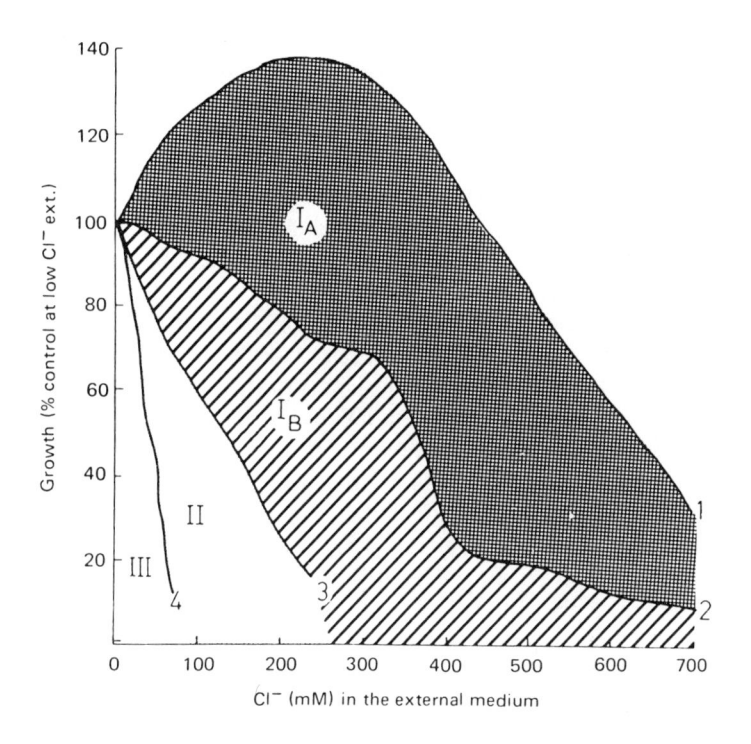

Many plant species occur naturally on saline soils though halophytic crops of importance include only a few pasture and herbage species. Mechanisms of salt tolerance vary greatly between these halophytic species, and include salt excretion, selective ion uptake and succulence (Flowers et al., 1977). Unlike salt tolerant prokaryotes, the isolated enzymes of salt tolerant higher plants seem to be just as sensitive to NaCl as the same enzymes from salt intolerant plants (Flowers et al., 1977; Greenway and Osmond, 1972). Tolerance appears to result from an ability to exclude NaCl from the sites of active metabolism and to balance the low osmotic potential of the vacuole by the production of organic osmotica in the cytoplasm, in particular proline, betaines, sorbitol, sucrose, maltose or rhamnose (Wyn Jones et al., 1977; Stewart et al., 1979; Wyn Jones, 1981; Briens and Larher, 1982).

6.2.3 Salinity and photosynthesis

Salinity at sub-lethal levels reduces productivity both by reducing leaf area and leaf photosynthetic rate (Gale, 1975; Jensen, 1975). The question of which one of these factors is the more important in determining crop photosynthesis cannot be answered from existing information. In theory salinity could affect photosynthesis by the three main routes outlined above. A high electrolyte concentration will depress the osmotic potential of the soil water and thus the water potential of the soil (Ψ_{soil}) producing the so-called "physiological drought" which was once thought to be the basis of salinity-induced reduction of plant production (Strogonov, 1962). Since electrolytes have a more detrimental effect than equiosmolar concentrations of organic osmotica (e.g. mannitol) a lowered plant water potential (Ψ_{plant}) is not the sole effect of salinity on productivity. For plants to obtain water from soil, Ψ_{plant} must be lower than Ψ_{soil}. For example, one consequence of a soil water salinity level of 200 mM is a Ψ_{soil} of ≤ -1.0 MPa. Ψ_{plant} will therefore always be less than -1.0 MPa for the plant to avoid desiccation. This is sufficiently low in mesophytes to inhibit processes related to photosynthesis, including stomatal opening, chlorophyll synthesis, and Ru*bis*CO synthesis (Hsiao, 1973). Ψ_{plant} is further depressed in plants growing in saline conditions by salt induced reduction of the soil hydraulic conductivity which increases the resistance of water movement from the soil to the root surface.

Salinity has been shown to induce nutrient deficiencies in soils which would otherwise be considered to have an adequate supply of those nutrients. Uptake of K^+, NH_4^+, NO_3^-, Mg^{2+} and $Fe^{3+/2+}$ are inhibited by NaCl (Solov'ev, 1969 a,b). Salinity-induced deficiencies of K^+ and Mg^{2+}, in particular, could be expected to depress photosynthesis because of their vital roles in stomatal opening and thylakoid-stroma ion gradients, respectively, although one report has shown that Cl^- concentration up to 500 mM enhances PSII electron transport in chloroplast membranes of the Mangrove *Rhizophora mucronata* (Critchley, 1982). Finally, salts are directly toxic to a number of vital processes inside the cell. NaCl *in vitro* disrupts the tertiary structure of enzymes and membranes (Lapina and Bikhukhometova, 1972), protein synthesis and oxidative phosphorylation (Flowers *et al.*, 1977) at osmotic potentials which are not in themselves inhibitory.

In practice salinity reduces the photosynthetic influx of CO_2 (F) at rooting medium salinities of ≤ 150 mM and in many crops this decline in F is initially a result of partial stomatal closure with increase in salinity. The causes of closure can apparently vary between species. In *Allium cepa* stomatal closure occurred following loss of turgor in the leaf whilst in bean (*Phaseolus vulgaris*) closure occurred despite high leaf turgor. Reductions in F in response to salinity are not always a result of stomatal closure. In cotton (*Gossypium hirsutum*) despite a constant stomatal aperture, a decline in F occurred and was greater at higher CO_2 concentrations suggesting an effect on the biochemical pathways of CO_2 assimilation. Salinity *in vivo* has been shown to disrupt the ultrastructure of chloroplasts and to alter the fluorescence induction kinetics of photosystem II (Flowers *et al.*, 1977; Baker, 1978; Dominy and Baker, 1980) though there is no evidence that either change actually correlates with a change in F. However, it has been shown that salinity induced anatomical changes in leaves of bean (*P.*

vulgaris) and cotton (*G. hirsutum*) partially compensate for the increased resistance to CO_2 uptake (Longstreth and Nobel, 1979).

In conclusion, salinity decreases crop productivity through reduction of leaf area and perhaps by increased respiration (Gale, 1975; Hasson *et al.*, 1983). It is not clear how important reduced leaf photosynthetic rates are to decreased production at the whole crop level. To some extent the productive potential of the saline soils could be enhanced by exploring the available halophytic flora as potential crops (Epstein, 1972). Even if suitable food crops cannot be obtained, areas such as those salinized by irrigation in the Punjab and Maharashtra and now barren, could at least be used for fuelwood crops, since many halophytic shrub and tree species are known.

6.3 NITROGEN AND NUTRIENTS

6.3.1 Nitrogen and photosynthesis

Nitrogen is the basic constituent of amino acids and as the production of protein is directly proportional to the availability of nitrogen, this element plays an important role in biomass production (Arckoll and Festenstein, 1971; Cooke, 1975). Nitrogen is a component of chlorophyll, the intermediates of the chloroplast electron transport chain, and the enzymes of the dark reactions. R*ubis*CO alone, may account for half of the total leaf protein in C_3 species. Application of nitrogen to plants can lead to dramatic increases in leaf area. Nitrogen is thus of obvious importance to photosynthetic production at several levels. The element is taken up by non-leguminous crops predominantly as ammonium (NH_4^+) or nitrate (NO_3^-) ions and is supplemented in legume crops by the direct fixation of atmospheric nitrogen into ammonia by nitrogenase. Energy is required for the active uptake of NO_3^- and NH_4^+ into the roots and the incorporation of nitrogen into organic molecules. As energy is also required for the conversion of nitrate to ammonia, the energetic requirements are theoretically greater in terms of carbohydate for producing proteins from NO_3^- than from NH_4^+ but yields of dry matter are in general similar (Penning de Vries *et al.*, 1974; Bledsoe, 1976; Miflin, 1980). It appears that the derivation of energy for nitrogen metabolism directly from photosystem I may explain this anomaly. Several steps in the synthesis of glutamate from nitrite can be powered by light energy within the chloroplast (Anderson and Done, 1977a,b; Lea and Miflin, 1979; Guerrero *et al.*, 1981). The utilization of nitrogen is therefore associated with components of photosynthesis but whereas much information is available on nitrogen metabolism and amino acid biosynthesis (Miflin, 1980) considerably less is known about the effects of nitrogen and other nutrients on photosynthesis.

Net photosynthesis increased linearly with leaf nitrogen content in a C_4, C_3 and "intermediate" species though the photosynthetic rate of the C_4 species was twice that of the C_3 and "intermediate" at similar N contents above 2% of dry matter (Bolton and

Brown, 1980). In C_4 plants the quantity of C assimilated per unit of nitrogen in the plant is approximately twice that in C_3 plants (Brown, 1978; Osmond et al., 1982). The presence of nitrogen is critical even at the seedling stage. More rapid leaf emergence and higher photosynthetic rates per plant were observed in *Hordeum sativum* cultivars which had higher grain nitrogen contents whilst delayed application of exogenous nitrogen reduced the content of Ru*bis*CO (Metevier and Dale, 1977). In a *Triticum aestivum* crop, the photosynthetic rates of the flag leaf and second leaf decreased because of increased self shading in the dense canopy of a high nitrogen treatment but the productivity of the flag leaf increased due to the increase in area (Pearman *et al*, 1979). The photosynthetic CO_2 assimilation of eight day-old barley seedlings supplied with 5 mM NO_3^- was 25 times greater than a leaf receiving 1 mM NO_3^- and this was associated with marked differences in nitrate reductase activity in the two treatments (Morrison et al., 1979). Nitrate therefore stimulates *de novo* synthesis of the enzyme used for its own reduction.

Leaf discs of *Medicago sativum* photosynthesized at higher rates in the presence of NO_3^- or NH_4^+ than in controls and there was an increase in the concentration of alanine and some other amino acids. Since there was a concurrent reduction of pyruvate and increase in phosphoenolpyruvate, Plaut *et al.* (1976) suggested that NH_4^+ stimulated the activity of pyruvate kinase and accelerated the rate of transfer of photosynthetically incorporated carbon into the synthesis of α-keto skeletons for amino acid synthesis. The effect of N on incorporation of carbon within and between C_3 and C_4 plants differed. NO_3^- and NH_4^+ decreased the proportion of carbon fixed into sucrose and starch and increased that fixed into malate and aspartate in *Zea mays* (a malate forming C_4 species, Blackwood and Miflin, 1976). There was also a higher aspartate : malate ratio with increasing NH_4^+ in *Zea mays* (Tew et al., 1975). In contrast, there was no effect of NH_4^+ on the carboxylation productions of *Eragrostis curvula* (an aspartate forming C_4 species) or in *Hordeum sativum* (C_3). The formation of well-developed grana in the bundle sheath cells of C_4 plants was decreased by NO_3^- and stimulated by NH_4^+, and with the exception of *Zea mays*, the CO_2 compensation point increased and net photosynthesis decreased with increasing concentration of NH_4^+ (Tew et al., 1975). These responses are clearly complex and not well defined at present.

Deficiencies of nearly all the essential nutrients reduce the photosynthetic rate of higher plants (Bottrill *et al*, 1970) and nitrogen deficiencies often result in lower levels of photosynthetic enzymes (Wong, 1979a). A shortage of N and K but not P decreased shoot dry weight and net photosynthesis in the first leaf of *Hordeum sativum* during a four week experimental period (Natr, 1970). The removal of P and K from nutrient solutions caused substantial reductions (66%) in net photosynthesis and increased mesophyll and stomatal resistances after 30 and 21 days respectively of *Beta vulgaris* (Terry and Ulrich, 1973a,b). Inhibition of NADP reduction was substantially correlated with the photosynthetic activity in *B. vulgaris* but photophosphorylation was only correlated with the level of available phosphorus (Tombesi et al., 1969). Large reductions in net photosynthesis of *Gossypium hirsutum* were obtained at low concentrations of N, P or K and were related to a similar increase in mesophyll resistance per unit of cell wall area. The stomatal resistance remained constant (Longstreth and Nobel, 1980). Phosphorus deficiency also diminished protein relative to carbohydrate synthesis in *Nicotiana tabacum* (Katie, 1970) and deficiencies of all essential nutrients except iron caused reduced Hill reaction activities in *Lycopersicon esculentum* and *Spinacea oleracea* (Spencer and Possingham, 1960). Some other effects of nutrient deficiency on chloroplast activity are considered by Possingham (1970).

6.3.2 Nitrogen fixation

A number of organisms can fix nitrogen through symbiotic associations and of particular interest are the Leguminosae which include several important crops. Biological nitrogen fixation world-wide remains the major source of inorganic nitrogen for plants. A high energy requirement from respiratory activity is needed for reduction of molecular nitrogen as well as the growth and maintenance of the microbial symbiont. Symbiotic nitrogen fixation is dependent on an adequate supply of photosynthate for respiration if the nitrogen fixing tissues are to efficiently utilize available substrate. It can be considered that the efficiency of symbiotic nitrogen fixation is related to the photosynthetic characteristics of legume crops and the two processes modulate each other (Bethlenfalvay *et al.*, 1978a,b; Lamborg, 1980).

Minchin and Pate (1973) have reported that as much as 32% of the total photosynthate is translocated to the nodules where 5% is used for nodule growth, 12% in respiration and 15% returned to the shoot via the xylem tissue. The theoretical minimum requirement for the nitrogenase reaction is approximately 2 kg of carbon used for every kg of nitrogen fixed (Bulen and Le Compte, 1966; Hardy and Havelka, 1976), approximately half of that required for nodule fixation in cowpea and white clover (Haystead and Sprent, 1981; Ryle *et al.*, 1979). Published estimates in general suggest that the total cost of nitrogen fixation in legumes (including nodule growth and maintenance) can vary between 3.8–6.4 kg of carbon used per kg of nitrogen fixed (see Hardy *et al.*, 1978).

Reduction of molecular nitrogen also requires the hydrolysis of a minimum of 6 molecules of ATP for each NH_3 formed but these minimum requirements may be exceeded due to the inefficiency of the nitrogenase systems and in some legumes due to the occurrence of hydrogenase activity (Haystead and Sprent, 1981). It would appear therefore that nitrogen fixation is an energetically costly process and although much nitrate reduction may occur at little energy cost to the plant, nitrogen reduction by legumes may be limited by the supply of photosynthate under field conditions (Hardy and Havelka, 1976; Watt *et al.*, 1975; Anderson *et al.*, 1977; Ursino *et al.*, 1979; Miflin, 1980).

The pattern of nitrogen fixation and growth varies considerably. For example, nitrogen fixation rose to a maximum in peas just before flowering and then declined slowly during pod fill (La Rue and Kurz, 1973). This observation can probably be explained by the avilability of assimilates for nitrogen fixation since the amount of photosynthate translocated to nodules during the vegetative phase in *Glycine max* was greater than that translocated during the reproductive period when the demands for nitrogen and carbon during pod fill were much greater (Latimore et al., 1977). A major feature of nitrogen fixation is its decreased activity in the presence of NO_3^- or NH_4^+ ions. Supply of these ions to the roots of *Glycine max* decreased the amount of carbon in nodules (Latimore et al., 1977; Ursino et al., 1979). The effect of NO_3^- was greater than that of NH_4^+ as the reduction of NO_3^- probably competed with the nodule for photosynthate though there may be other explanations for this effect (Haystead and Sprent, 1981). In another experiment, 5 mM NO_3^- treatment of *Lupinus* promoted similar rates of growth as in symbiotic plants in spite of a lower efficiency of conversion (57% vs. 69%) and higher CO_2 loss per unit of nitrogen (10.2 vs. 8.1 mg C/mg N) in the nodulated plants (Pate et al., 1979). The growth of legume crops therefore does not appear to be limited by the high energy requirements of symbiotic nitrogen fixation. The efficiency of symbiotic nitrogen fixation has been reviewed by Philips (1980).

Nitrogen fixation is energeticaly an "expensive" process. This has been demonstrated by a substantial increase in nitrogen fixation when photosynthetic capacity has been increased through CO_2 enrichment of canopies of field grown *Glycine max, Arachis hypogea* and *Pisum sativum* (Watt et al., 1975; Anderson et al., 1977; Hardy and Havelka, 1976). There is also evidence that some tropical C_4 grasses are able to fix nitrogen through associative symbiosis with *Azospirillum* spp. and that the coastal C_4 grass *Spartina alterniflora* has a symbiotic association with a nitrogen-fixing bacterium (Döbereiner et al., 1978; Patriquin, 1978; Ela et al., 1982). Because of their greater capacity to fix CO_4, C_4 grass species are likely to provide more photosynthate for growth and nitrogen fixation though the potential success of this symbiosis may be limited by factors other than photosynthate supply.

The reduction of nitrogen by the nitrogenase system in the root nodule also gives rise to the production of hydrogen from protons. As this requires energy in the form of ATP, it can be considered a waste of energy as far as nitrogen fixation is concerned (Haystead and Sprent, 1981). To offset this loss, the nodules may also possess a hydrogenase system which can be oxidized to produce ATP, thus recycling the energy used in hydrogen formation. The activity of the hydrogenase system however differs between strains of *Rhizobium*. For example, when *Vigna unguiculata* are inoculated with *Rhizobium* strain 32HI, H_2 evolution is not detected. Likewise, soybeans inoculated with USDA 110 produce nodules which evolve negligible amounts of H_2 in air (Eisbrenner and Evans, 1983). In these instances therefore, the hydrogenase system is very efficient. However, nodules obtained from *Medicago sativa* and *Trifolium* spp. plants inoculated with selected strains of *Rhizobium* are relatively active in H_2 evolution and thus show a net loss of energy.

CHAPTER 7
Pollution

7.1 POLLUTANTS IN THE ENVIRONMENT

The precise consequences of pollutants on productivity and photosynthesis are not well-defined, and no cohesive theory has yet emerged from the voluminous literature to relate cause and effect. Their effect on dry matter production will in general be a function of the duration of exposure and concentration of pollutant stress, or alternatively the rate of uptake of the pollutant. Species differ in their sensitivity to exposure to pollutants however and final yield losses will be determined by whether the stress is reversible, chronic or lethal (Larcher, 1981). Even if no loss in production is incurred, the external appearance or quality of crops may be so affected by pollutants as to make them unmarketable (Weinstein and McCune, 1979).

Some pollutants occur naturally at low levels in the environment and plants are adapted or may even benefit from their presence. Levels of pollutants are enhanced from atropogenic sources (e.g. SO_2) and others are entirely anthropogenic in origin (e.g. HF). For example, sulphur dioxide contains an element which is essential to plants and of potential benefit to both photosynthesis and yield, at least in small quantities (Lougham, 1964; Cowling et al., 1973; Müller et al., 1979). Others, e.g. zinc, are essential micronutrients but may become highly toxic to plants at elevated levels (Hampp et al., 1976). Carbon dioxide is a special case. It is essential for photosynthesis, but owing to the burning of fossil fuels and possibly deforestation, it is now increasing in concentration to the extent that it may be classed as a potential pollutant although overall effects of this increase on global vegetation are as yet unclear.

Over the last 30 years, pollutants have become a particular problem. Locally, marked pollution occurs around many industrial sites, especially smelters, but more general pollution results from the burning of fossil fuels and in particular because of heavy automobile traffic. This has prompted much research into their effects on photosynthesis and production (Unsworth and Ormrod, 1982). In addition, pollutants may have secondary effects, most notably interference with stratospheric ozone concentrations leading to increased terrestrial receipts of ultra-violet radiation. Thus, studies of the effects of ultra-violet radiation may also be classed as studies of pollution effects.

A whole spectrum of pollutants, heavy metals, the oxides of nitrogen and sulphur, ozone and ultra-violet radiation have been studied in relation to photosynthetic processes, though sulphur dioxide has received by far the most attention. The photosynthetic measurements which have been made to date need careful interpretation. Meaningful results are difficult to obtain as the concentration of pollutants often varies temporally and haphazardly under field conditions and the approaches which laboratories have used to simulate these conditions differ. As in water stress experiments, gross changes observed in whole plants *in vivo* are further studied at the chloroplast level *in vitro*, but similarly the use of heterogeneous populations of isolated chloroplasts which have partly or wholly lost their envelopes may give spurious results (Hällgren, 1978; Heath, 1980).

7.2 GASEOUS POLLUTANTS

7.2.1 Carbon dioxide

Earth is habitable only because of the 300 ppm CO_2 in its atmosphere. This level of concentration is believed to have been reached after periods of fluctuations during the 3,400 million years of the evolution of photosynthesizing organisms (Reimer *et al.*, 1979; Barghoorn, 1984). If this concentration were much lower, life would have been impossible on the planet because it would have been too cold for photosynthetic activity. Over geologic history, a perfected system for the distribution of CO_2 has evolved, the carbon cycle fluxing between the atmosphere, vegetation, oceans and the rocks (Bolin *et al.*, 1979; Arthur, 1982; Pollack, 1982). Under natural processes, the atmosphere would continue to receive CO_2 from volcanic activity and hydrothermal activity which at the present rate of release would double the atmospheric carbon in 0.4 million years (Holland, 1978). However, in the last 100 years the global CO_2 concentration has increased to the present 340 ppm and further increases are expected to lead to a doubling or even a several-fold increase during the next one or two centuries. Apart from the expected modification of the global climate (Bolin, 1981), these unprecedented increases in CO_2 concentration are expected to affect plant productivity. Just how, and the implications of such effects, are as yet unclear. If the scenario which predicts a climate deterioration and as yet undetermined serious effect on plant productivity holds, CO_2 would indeed be considered a significant atmospheric pollutant.

Diurnal and seasonal fluctuations in carbon dioxide concentration of 5–15 cm^3 m^{-3} which result from photosynthetic and respiratory exchanges between vegetation and the atmosphere are of little consequence to productivity. The long-term increase in atmospheric CO_2 concentration however has prompted experiments to measure its effect on photosynthesis and productivity. For example, Enoch (1977) has calculated that the mean yearly photosynthesis of a C_3 species in

Israel has increased by 3.1% since the beginning of this century.

An atmospheric composition similar to the present was reached between 600 and 900 million years ago. But, there is no precise information on the CO_2 concentration of the atmosphere 100-150 years ago, although it is believed that it remained at the level of 280 ± 20 from 10,000 years ago until the beginning of this century (Watts, 1982).

It was estimated to be 290 cm^3 m^{-3} in 1860 (Bray, 1959) somewhat higher than the value of 268 cm^3 m^{-3} for 1850 derived from tree ring studies (Stuiver, 1978). The latter figure is more consistent with Brown and Escombe's (1905) measured value of 274 ± 5 cm^3 m^{-3}. By 1974, mean atmospheric concentration was 330 cm^3 m^{-3} (Baes et al., 1976). Current usage of fossil fuels at the rate of 1.8×10^9 tonnes per annum should increase the atmospheric CO_2 concentration by 2.4 cm^3 m^{-3} per annum but at present, the annual increase is 0.7 cm^3 m^{-3} (Bolin et al., 1979). Current projections suggest an approximate doubling of the 1860 level by 2030, and Bacastow (1981) suggests a concentration of 2500 cm^3 m^{-3} by the year 2100. Whatever the exact level, the picture depends on which CO_2 release scenarios (Perry, 1982) and which carbon cycle models (Bolin, 1981; Arthur, 1982) are used to simulate them.

The net change in CO_2 concentration depends on the rate of emission of CO_2 and the absorption of excess production into carbon reservoirs, particularly the surface thermocline gyre waters and the deep sea. The absorption capacity of these reservoirs has been discussed by Kerr (1980). Besides the major source of CO_2, viz. the burning of fossil fuels, an additional input from the destruction of forests is thought to occur (Freyer, 1979).

Terrestrial vegetation and dead organic matter in the soil constitute significant components of the carbon cycle both in size and time scales of their turnover (Whittaker and Likens, 1975; Schlesinger, 1977; Bolin et al., 1979). However, because of significant gaps in knowledge relating to effects in global primary productivity due to increasing atmospheric CO_2 concentrations and the scanty historical qualitative data on anthropogenic sources of CO_2, the general features of the terrestrial carbon cyle are only painted in broad outlines. Simulation equations, notably those of Moore et al., (1981) and Bolin et al., (1981) suggest a simple model for the analysis of the role of terrestrial ecosystems in the global carbon budget which could serve as a structural framework for biome modelling. These divide the biosphere into carbon pools. Such a structural framework permits both the incorporation of new data and experimentation with various hypotheses relating the rate of photosynthesis to atmospheric CO_2 or nutrient availability and also the role of respiration and bacterial decomposition in the soil.

Two recent publications (Lemon, 1984; Enoch and Kimball, 1984) have comprehensively treated the current knowledge of plant reactions to elevated CO_2. The two obvious effects of CO_2 enrichment are an increase in net photosynthesis and a decrease in transpiration rates. In controlled experiments, all plants (C_3, C_4 and CAM) subjected to an atmosphere with 21% O_2 benefit from addition CO_2 above the ambient 340 ppm (Zelitch, 1982). From these controlled environment experimental situations, some steps are being taken to move to naturally occurring vegetation and to reactions of single crops and of ecosystems. The day is still far off where all uncertainties would have been resolved.

Bazzaz (1980) suggested that plant competition would alter species composition significantly but that productivity would in general increase, particularly as a result of increased water-use efficiency. This would result from the increased size of the gradient of CO_2 concentration between the atmosphere and site of carboxylation within the leaf i.e. for a given leaf resistance to CO_2 diffusion an increased rate of CO_2 influx would occur whilst transpiration would be unaffected. In contrast, Goudriaan and Ajtay (1979) considered that any increase in CO_2, except under optimum growth conditions, would play a secondary role in the environment, and productivity would be primarily limited by the shortage of water and nutrients. Increased productivity could have important implications in natural ecosystems which have evolved complex food webs and competitive balances to fit a constant food supply. Predictions of increased productivity however are based on studies of plants grown in conditions where water and nutrients are not limiting. As this is rarely the case in natural ecosystems, it is possible that these limitations would interact with increased CO_2 concentrations and might result in little or no effect on productivity.

Present knowledge of the physiology of carbon assimilation suggests that photosynthesis will increase with CO_2 concentration more in C_3 than in C_4 plants (see Wong, 1979a; Wittwer, 1982). Patterson and Flint (1980) have recently investigated the effects of various concentrations of CO_2 on growth and biomass production in C_3 and C_4 species raised in environmental chambers. They found that high levels of atmospheric CO_2 increased the dry matter production in C_3 species but had negligible effect on C_4 plants. The authors concluded that atmospheric CO_2 enrichment will make C_3 crop plants more competitive with their C_4 counterparts.

Other investigators have attempted to raise the water-use efficiency and yield of C_3 plants by increasing atmospheric CO_2 concentration in growth chambers/glasshouses (Tinus, 1974; Gifford, 1979). Hardy and Havelka (1976) have obtained higher yields of field grown legumes by enriching their canopies with CO_2, but it should be pointed out that none of the techniques hitherto described on CO_2 enrichment of canopies of field grown crops is of any practical agronomic use. It is also unlikely that any technological solution to the problem of providing agricultural crops with higher CO_2 concentrations would have any significant effect on biomass production and economic yield in C_4 species.

Further study of the role of CO_2 in plant growth and development is required however if we are to understand the effect of elevated levels on productivity since plant responses to CO_2 appear to depend on complex interactions which are not readily predictable (Wong, 1979a; Rogers et al., 1980). Clearly such studies should bear in mind the suggestion that increased concentration may have profound consequences in changing the earth's climate. These changes relate to the trapping of outgoing long-wave radiation, the so called greenhouse effect, causing projected increases in mean global temperatures (up to 3°C for a doubling of atmospheric CO_2 concentration) and alteration in rainfall patterns. Current estimates of these effects are in themselves still speculative.

7.2.2 Sulphur dioxide

Sulphur dioxide is the most abundant sulphur-containing pollutant. The emissions come mainly from fossil fuel consumption and in the industrial countries of the world concentrations of 20 to 50 µg kg^{-1} occur over large areas of agricultural land (Fowler and Cape, 1982). These man-made emissions have increased considerably in recent decades in both Europe and North America. Since 1950, SO_2 emission from North Western Europe has doubled (Overrein et al., 1980), calculated at about 33 million tons in Europe (Grennfelt, 1981), and in the United States there has been a 26% increase since 1940, attributable mainly to greater electricity usage (Anon 1982). In Canada aggregate SO_2 emissions have fluctuated between 4.5 million metric tons in 1955 to 6.6 metric tons in 1965 largely as a result of changes in emissions from copper and nickel smelters. SO_2 pollution is supplemented by particulate SO_4, acid rain and H_2S. There appears to be a threshold level of SO_2 concentration below which no damage occurs. To some extent this threshold is a function of the sulphur status of the soil above which yield reductions occur particularly in certain crops (Anon, 1978).

The scope and nature of possible environmental impacts of SO_2 are only beginning to be understood, and the matter is further complicated because it is difficult to separate pollution impacts from the array of other stresses in the natural environment. Researchers in Central Europe and Eastern North America (Tomlinson and Silversides, 1982) report that spruce trees in some areas suffer "crown die-back" where leaves or needles at the treetop turn yellow, then brown, and ultimately drop off. Studies in several Central European countries, where the sulphur deposition levels are extremely high reveal serious forest vegetation problems. Whereas the negative effects of SO_2 and other acid rain producing pollutants affect forests through a number of pathways, those that relate to leaves and other photosynthesising plant organs would have the most direct consequences on plant productivity.

Much of the information concerning the effects of SO_2 on photosynthesis and dry matter production appears contradictory and inconsistent (Jeffree, 1976). Many of the earlier studies into the relationship between photosynthesis and SO_2 used unrealistic concentrations of SO_2 (>1 µg kg^{-1}) which greatly exceeded the levels of exposure which crops are likely to encounter in the field. Results from these experiments should be treated with caution. There is considerable evidence to suggest that the rate of photosynthesis of a number of species is inhibited by sulphur dioxide fumigation (Fig. 7.1; Bennett and Hill, 1973; Bull and Mansfield, 1974; Black and Unsworth, 1979a,b; Koziol and Jordan, 1979). Although photosynthetic rates of CO_2 assimilation by shoots following SO_2 fumigations have been observed these enhancements may be temporary or misleading (Black and Unsworth, 1979b; Winner and Mooney, 1980c).

The plant cuticle forms an effective barrier to the entry of SO_2 into the plant so that the entry would appear to be limited to the stomata (Bonte, 1975). It is difficult to determine the precise role of stomata in the inhibition of photosynthesis by SO_2. In some species (*Lolium*, Cowling and Koziol, 1979; *Populus*, Noble and Jensen,

Fig. 7.1 The effects of SO_2 concentration (ppm) on net photosynthesis (% depression) in several species. □ Pisum sativum *(after Bull & Mansfield, 1974);* ■ Vicia faba *(after Black & Unsworth, 1979b);* ○ Oryza sativa *(after Taniyama, 1972);* ▲ Hordeum sativum *(after Biermett & Hill, 1973;)* △ Diplacus aurantiacus *(after Winner & Mooney, 1980a);* ● Heteromeles arbutifolia *(after Winner & Mooney, 1980b);* ▽ TRIPLEX SABULOSA *(after Winner & Mooney, 1980c) (reproduced with permission from Black, 1982).*

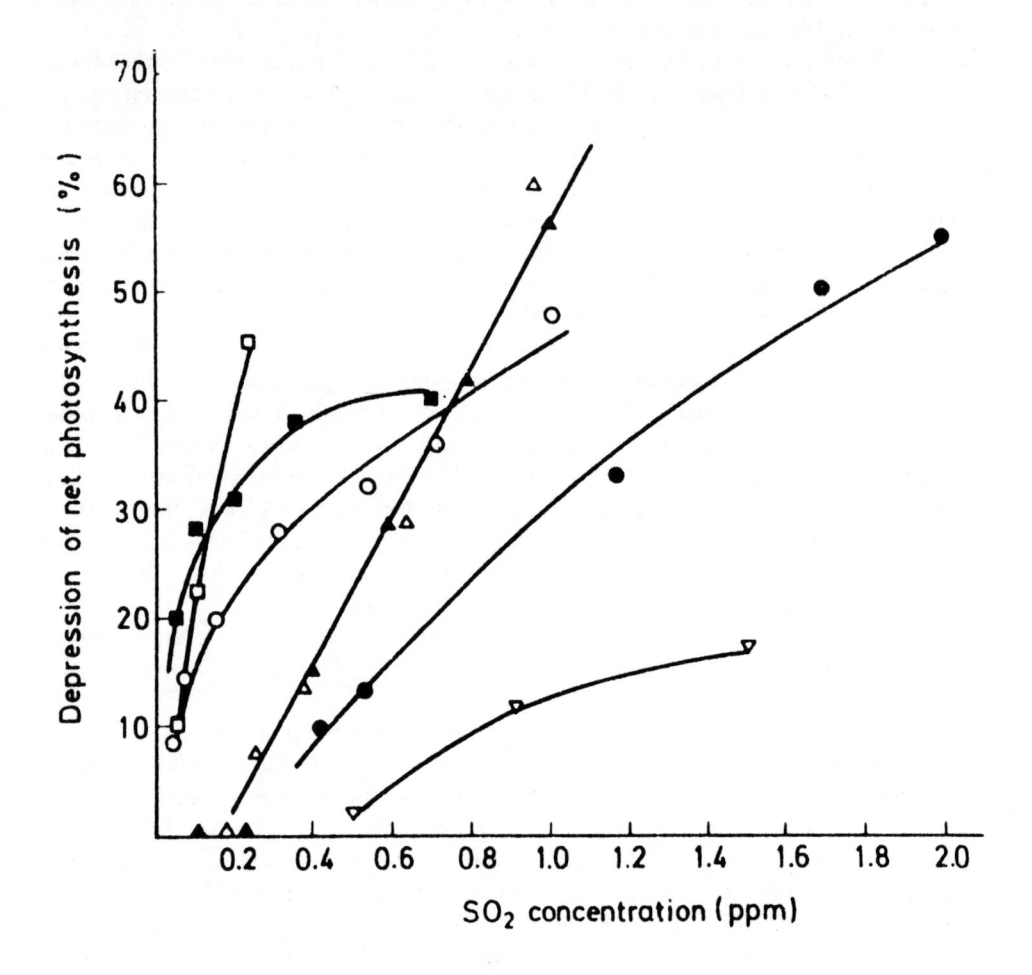

1979) stomatal resistance was possibly not affected as photosynthesis remained constant in the presence of SO_2. In contrast, photosynthesis increased at low but decreased at high concentrations of SO_2 in *Glycine max* (see also Coyne and Bingham, 1978 for similar results with H_2S) whereas stomatal resistance increased at both high and low concentrations though this stomatal response lagged behind the photosynthetic response (Müller *et al.*, 1979). In contrast, r_s was significantly lower in *Vicia faba* treated with $\geqslant 35$ g kg^{-1} SO_2 compared to the controls and photosynthesis was reduced (Black and Unsworth, 1979b), a common response when stomatal opening is observed and one in which the pollutant facilitates its rapid entry into the crop (Jeffree, 1976; Unsworth, 1981). From a survey of the literature Black (1982) concluded that enhanced stomatal opening usually occurred at moderate levels of SO_2 pollution, though the response is a function of VDP (Black and Unsworth, 1980).

A linear relationship between uptake and the total conductance (1/resistance) for SO_2 in *Vicia faba* has suggested that the use of a resistance analogue provides a valid approach for analysing the uptake of SO_2 and other gaseous pollutants by plants (Black and Unsworth, 1979c; Unsworth, 1982). In some instances however the SO_2 flux has been poorly correlated with stomatal conductance. For example, in Scots pine (*Pinus sylvestris*), if a shoot was darkened during the day an accelerated uptake of SO_2 occurred while r_s increased or remained unchanged (Hällgren, 1980). Although SO_2 can therefore induce a number of stomatal responses (Black, 1982) stomatal resistance may not be the primary factor in the regulation of photosynthesis by SO_2 or other pollutants. Stomatal responses to SO_2 are also independent of any action of SO_2 or CO_2 exchange (Unsworth, 1981). There is also evidence that unlike CO_2, the mesophyll may be a near infinite sink for SO_2 since gaseous concentrations of CO_2 in the substomatal cavities of *Vicia faba* were close to zero even during periods of rapid SO_2 uptake (Jeffree, 1976; Black and Unsworth, 1979c).

The deleterious effects of SO_2 are also manifested in the mesophyll cells where the gas is metabolized to sulphite, bisulphite and sulphate (Pucket *et al.*, 1973). These effects are often qualitatively similar to those associated with water and temperature stress. For example, reductions in productivity in several species after SO_2 treatment were associated with the loss of chlorophyll a and to a lesser extent chlorophyll b (Bortitz, 1964; Malhotra, 1977). In contrast, no changes were observed in the total chlorophyll content of *Populus* cuttings during fumigation with SO_2 (Jensen, 1975). In general however the rate of photosynthesis is normally affected before any change in chlorophyll content is observed (Hällgren, 1978).

Within the chloroplast SO_2 has been shown to affect electron transport, photophosphorylation, and the enzymes of the reductive pentose phosphate cycle (Ziegler, 1975; Hällgren, 1978). The permeability of the chloroplast membranes increases after SO_2 treatment. This is consistent with the leakage of primary photosynthetic products, inhibition of photophosphorylation and more recent evidence for the eventual peroxidation of membrane lipids in the presence of SO_2. (Fischer, 1973; Nobel and Wang, 1973; Coulson and Heath, 1974; Puckett *et al.*, 1974; Niebor *et al.*, 1976; Yu *et al.*, 1982). Ultrastructural studies suggest that reversible chloroplast swelling precedes visible symptoms of injury (Wellburn *et al.*, 1972; Godzik and Sassen, 1974; Thomson, 1975; Fischer *et al.*, 1976).

The fixation of bicarbonate by Ru*bis*CO was completely inhibited by SO_3^{2-} in *Spinacia oleracea* but not in *Pinus sylvestris* (Ziegler, 1972; Gezelius and Hällgren, 1980). Competitive inhibition may not be a general phenomenon of inhibition of photosynthesis by SO_2 as suggested by Ziegler (1975). PEP carboxylase was less sensitive at similar concentrations of the inhibitor SO_3^{2-} (Ziegler, 1973, 1974), and there is some evidence that C_4 plants are less sensitive than C_3 plants to SO_2 (Winner and Mooney, 1980c). Several other enzymes of carbon metabolism were also affected.

7.2.3 Nitrogen oxides (NO_x)

The oxides of nitrogen, NO and NO_2 are combustion products of fossil fuels. They are therefore components of vehicle exhaust emissions and also CO_2 enrichment systems in glasshouses which involve the burning of hydrocarbon fuels (Derwent and Stewart, 1973; Unsworth, 1981). Law and Mansfield (1982) have summarized the occurrence of NO_x pollution.

Capron and Mansfield (1976) have investigated the effects of NO and NO_2 (0.1–0.5 cm^3 m^{-3}) on net photosynthesis and found that both gases reduce photosynthesis to the same extent and that their effects are additive. Reductions in the productivity or photosynthesis of *Lycopersicon esculentum, Capsicum annuum* or *Triticum aestivum* have been observed after NO_x treatment (Spierings, 1971; Prasad and Rao, 1979; Law and Mansfield, 1982). Photosynthesis and transpiration of *Phaseolus vulgaris* were also inhibited by NO_2 (1.0 to 7.0 cm^3 m^{-3}) but since transpiration was less affected, Srivastava *et al*. (1975) suggested that the principal effects were produced through absorption of NO_2 by the mesophyll. It has been suggested that the major effect of NO_x is to saturate nitrite reductase and thereby cause a build up of toxic levels of nitrite (Capron and Mansfield, 1976).

7.2.4 Ozone (O_3)

There has been relatively little work on the effects of ozone on the photosynthetic performance of plants. This may be because this pollutant is more important in areas of high temperature and irradiance, since both are required for significant photochemical formation of O_3 from its precursors. Ozone is a reactive molecule with two unpaired electrons and in water can form hydroxyl, hydroperoxyl, superoxide and other free radicals (Peleg, 1976). As a result the primary effect of ozone is thought to be damage to cell membranes.

Stomata are probably the major pathway for the entry of ozone into plants but it is not yet clear whether the flux, and therefore to some extent the degree of injury, is inversely proportional to stomatal resistance (Tingey and Taylor, 1982). Photosynthesis was reduced in the presence of ozone and there was some differential sensitivity between species (Hill and Littlefield, 1969; Carlson, 1979). These reductions in photosynthetic CO_2 fixation were manifested in both the stomata where closure was observed and in the chloroplast through membrane damage (Table 7.1; Hill and Littlefield, 1969; Beckerson and Hofstra, 1979a,b). Ozone inhibited both PS I and PS II in isolated chloroplasts of *Spinacia oleracea* and *Nicotiana tabacum* but without uncoupling photophosphorylation (Chang and Heggestad, 1974; Coulson and Heath, 1974; Koiwai and Kisaki, 1976). A decline in the activity of Ru*bis*CO has also been observed in the presence of ozone (Nakamura and Saka, 1978).

Table 7.1   Effects of Ozone on Photosynthesis[1]

Species	Ozone (ppm)	Exposure (h)	Inhibition (%)
Avena sativa	0.4	0.5	33
Nicotiana tabacum	0.4	1.5	78
Lycopersicon esculentum	0.6	1.0	43
Phaseolus vulgaris	0.6	1.0	29
	0.3	3.0	22
Glycine max	0.4	4.0	37
	0.6	2.0	19
Medicago sativa	0.1	1.0	4
	0.2	1.0	10
Quercus velutina	0.5	8.0[2]	30
Acer saccharum	0.5	8.0[2]	21
Fraxinum americana	0.5	8.0[2]	0
Populus euramericana	0.9	1.5	50

[1] After Tingey and Taylor (1982).
[2] 4 hours per day for 2 days.

The ecological implications of atmospheric pollutants are complicated. The susceptibility of plants appears to be a function of the stage of plant development as well as soil and climatic factors and the adaptation of plants to pollutants varies within and between species (Jacobson and Hill, 1970; Weinstein and McCune, 1979). Some pollutants particularly SO_2 and NO_2, are often in the same atmosphere and it may be difficult in the field to isolate the individual effect of any one pollutant. Further studies particularly in the laboratory should therefore be carried out not only at realistic atmospheric levels of pollutants but also account for additive or synergistic reactions in the presence of more than one pollutant (e.g. Bull and Mansfield, 1974; Coyne and Bingham, 1978; Hampp et al., 1976; Lamoreaux and Chaney, 1978; Beckerson and Hofstra, 1979a,b; Noble and Jensen, 1979; Ormrod, 1982; Wellburn, 1982). Some of the effects of gaseous pollutants on the growth and productivity of crops are reviewed in Unsworth and Ormrod (1982).

7.3   HEAVY METALS

A number of studies have shown that the uptake of toxic levels of heavy metals inhibit photosynthesis. These include the accumulation of Cd, Ni, Pb, Tl, Zn, Co and Al (Bazzaz et al., 1974a,b; Huang et al., 1974; Schnabl and Ziegler, 1974;

Carlson *et al.*, 1975; Austenfeld, 1979; Van Assche *et al.*, 1979). Parallel falls in transpiration have suggested that this inhibition was primarily the result of stomatal closure (Bazzaz *et al.*, 1974a,b). Effects at stomatal level may be a secondary influence of heavy metal toxicity however, the primary sites of action being at the chloroplast level (Van Assche and Clijsters, 1983).

For example, Cd, Ni, Pb and Zn caused reversible or irreversible reductions in electron transport (Hampp *et al.*, 1976; Tripathy *et al.*, 1981). In the case of Zn this impairment occurred mainly at the oxidizing side of PSII and was associated with a decrease in phosphorylation capacity and NADPH production. Inhibition of Ru*bis*CO activity was also observed in the same experiment (Van Assche and Clijsters, 1983).

As with gaseous pollutants therefore, current evidence suggests that heavy metals appear to inhibit several parts of the photosynthetic process. Similarly, heavy metals are often present as mixed contaminants and additive or synergistic effects may be anticipated.

7.4 ULTRA-VIOLET RADIATION

UV (250–400 nm) is sub-divided into three wavebands. The middle waveband, UV-B (280-320 nm) is present in solar radiation at the earth's surface and causes biological damage, while UV-A (320–400 nm) activates repair mechanisms which alleviate the effects of UV-B. Levels of natural UV-B fluctuate with latitude through changes in solar angle and a natural gradient in the thickness of the stratospheric ozone layer, which is primarily responsible for the attenuation of UV radiation. Of major concern at the present time is a fear that the ozone layer is being depleted because of human activities, particularly from the release of chlorofluoromethanes leading to a chlorine-catalyzed reduction of ozone (Molina and Rowland, 1974). UV-B is the component of UV most sensitive to any change in the thickness of the ozone layer and enhanced levels will increase biological damage (Caldwell, 1981).

Action spectra suggest that the major receptor sites for UV-B include both nucleic acids and proteins, but the photosynthetic process is also involved at some level and rates of photosynthesis are depressed in the presence of UV-B (Caldwell, 1971; Sisson and Caldwell, 1977). Damage to the photosynthetic capacity of plants is related to the total accumulated exposure to UV-B. Visible and UV-A radiation protects plants from the deleterious effects of UV-B, though the precise nature of this phenomenon is not clear (Sisson and Caldwell, 1976; Teramura *et al.*, 1980; Caldwell, 1981). Photosynthesis would appear to be affected at the levels of primary photochemistry, electron transport, photophosphorylation and carboxylation (Okada *et al.*, 1976; Brandle *et al.*, 1977; Klein, 1978; Vu *et al.*, 1981). These observations suggest that the disruption of the thylakoid membrane and stromal lamellae may be ultimately responsible for the effects of UV-B in photosynthesis (Mantai *et al.*, 1970; Brandle *et al.*, 1977). Plants differ in their response to UV-B (Biggs *et al.*, 1975; National Academy of Sciences, 1979). Some species acclimate to the prevailing levels of UV-B by synthesizing flavenoids in their epidermal cells to absorb UV-B, thereby screening the underlying photosynthetic tissues: in some species alkaloids fulfil this function (Wellman, 1974; Levin, 1976; Robberecht *et al.*, 1980). C_4 plants may be more tolerant of UV-B than C_3 plants (Van and Garrard, 1976; Van *et al.*, 1976). The significance of current and future levels of ozone depletion may therefore be more in species distribution than in significant reductions of bioproductivity (Montfort, 1950; Caldwell, 1977a; Fox and Caldwell, 1978). The major effects of UV radiation in plants have been reviewed by Caldwell (1977b, 1981) and Klein (1978).

Part III
Research into Photosynthetic Productivity

Introduction

There has been a proliferation of research into photosynthetic productivity at all levels of organisation in the plant over the past thirty years. These studies encompass photochemistry and carbon metabolism at a chloroplast level and more recently at the leaf level; gas exchange of single leaves, whole plants, and plant canopies; and biomass production at field level. It is however unusual to find research programmes which have forged any close link between say the behaviour of a plant at the photochemical level (10^{-13}–10^{-9} s) with biomass production (10^7 s).

This section summarizes the present state of each research area and how their understanding has contributed to our knowledge of the relationship between photosynthesis and production. An integrative approach is taken starting with photochemistry and ending with biomass. Chapter 11 considers some of the models which have been developed to predict biomass production from our knowledge of photosynthesis both at a metabolic and environmental level.

CHAPTER 8

Photochemistry and Carbon Metabolism

8.1 PHOTOCHEMISTRY, ELECTRON TRANSPORT AND PHOTOPHOSPHORYLATION

The capture of photons by the chloroplast and their utilization in chloroplast electron transport forms the largest single area of current research in photosynthesis. Several reviews on light harvesting (Junge, 1977; Knox, 1977; Seely, 1977; Butler, 1978), chloroplast electron transport and photophosphorylation (Hall, 1976; Goldbeck et al., 1977; Jagendorf, 1977; Velthuys, 1980; Barber, 1982; Malkin, 1982; Barber, 1983; Haehnel, 1984) as well as light reactions in general (Trebst and Avron, 1977; Govindjee, 1983) summarize the current state of knowledge in these areas.

The light reactions of photosynthesis are located in the thylakoids and stromal lamellae of the chloroplast. Use of the freeze fracture technique for examination of thylakoids in the electron microscope shows the existence of particles in the membrane corresponding in size and number to the photosystem units suggested by spectroscopy (Sane, 1977; Mühlethaler, 1977). In green plants each unit consists of the reaction-centre chlorophyll which constitute less than 1% of total chlorophyll. The reaction-centre of photosystem I (PSI) consists of a dimer of chlorophyll a (Ballschmitter and Katz, 1972) complexed to a protein, and denoted P700-CPa (Thornber, 1975). An undesignated chlorophyll-protein has been ascribed to photosystem II and was identified from fluorescence emission spectra (Kitajima and Butler, 1975). Reaction centres are surrounded by antennae chlorophyll-a proteins which do not participate in photochemical reactions, but funnel excitons (energy derived from photon capture) to the reaction centres. In addition a light-harvesting chlorophyll a-b protein complex (LHCP) may transfer excitons to the antennae of PSII or PSI, or transfer excitons from the antennae of PSII to PSI. It is suggested that in darkness LHCP is in contact with PSII, but not PSI; this is termed "State 1". In a dark-light transition LHCP may initially transfer trapped light energy only to PSII. However, in the light LHCP is phosphorylated by ATP and this is associated with an increased ability to transfer light energy to PSI (Bennett et al., 1980). It is believed that this phosphorylation increases physical contact between PSI and LHCP particles in the membrane, so increasing the proportion of trapped light available to PSI. This change in distribution is known as a "State 1-State 2" transition which may be reversed in the laboratory by providing additional light in wavelengths absorbed only by PSI (Barber, 1982; Haworth et al., 1982). This mechanism provides a means by which the plants may balance the energy directed to PSII and PSI and so vary the ratios of non-cyclic to cyclic photophosphorylation. Presumably such a mechanism may ensure the most efficient use of trapped light depending on relative demands for ATP and chloroplast reductants. However, this mechanism is probably not essential to plants since mutants of *Hordeum sativum* lacking LHCP and a capacity to undergo a "State 1-State 2" transition are apparently able to grow and reproduce normally. Contrary evidence also suggests that PSI does not receive excitation energy from LHCP particles associated with PSII (Anderson and Melis, 1983).

Secondary pigment molecules are also present and include the carotenoids and α-tocopherol (vitamin E). These may assume the role of a sink for the degradation of excess energy which cannot be used by the electron transport system (Krinsky, 1971). The production of the reactive superoxide anion, O_2^-, by the transfer of an electron from chlorophyll to oxygen is therefore avoided. In this process, molecular oxygen may act as an electron acceptor, for the PSI electron transport chain at the iron-sulphur centres of the PSI and for reduced ferredoxin, producing the superoxide. Generation of superoxide is favoured by high O_2 and reduction of electron transport intermediates between PSI and NADP, as would result during, for example, decreased rates of carbon metabolism at low temperature. Superoxide may then be converted to hydrogen peroxide by a dismutation reaction. Hydrogen peroxide is toxic to CO_2 assimilation, but in the presence of superoxide may give rise to the extremely reactive hydroxyl free radical i.e. ·OH. This radical is extremely destructive to photosynthetic membranes and may cause photoinhibition, probably through an effect on or near to the PSII reaction centre (Satoh, 1970; Satoh and Fork, 1982). However, chloroplasts contain high levels of superoxide dismutases and peroxidases which will detoxify these damaging derivatives of oxygen reduction as well as large quantities of ascorbate and glutathione which will remove ·OH (Foyer and Hall, 1980; Halliwell, 1982a,b). The synthesis of these detoxification agents and carotenoids may be crucial to the protection of chlorophyll-protein complexes during periods of stress.

The balance between thylakoids and stromal lamellae, and the presence or absence of photosynthetic components on these membranes appears to be related to function. For example, ultrastructural studies suggest that only the stromal lamellae and exposed surfaces of the thylakoids possess the CF_1 required for photophosphorylation (Anderson, 1982; Barber, 1982). In NADP-me type C_4 plants, grana and PSII are largely absent from the bundle sheath chloroplasts (Laetsch, 1969). This can be related to the ability of these plants to generate a part of their reducing power during the decarboxylation of malate.

Cyclic electron flow could provide a "safety-valve" or alternative source of ATP when non-cyclic flow is inhibited. For example, under stress conditions when stomata close, the CO_2 supply diminishes and non-cyclic flow could decrease due to a feed-back of accumulated NADPH and reduced ferredoxin (FdH). In this instance, cyclic electron flow would maintain the supply of ATP for other cellular processes. It has already been shown to support photoassimilation of glucose and K^+ influx (Arnon, 1977). Cyclic electron flow is also reputed to be particularly active and important during the induction of photosynthesis from a darkened state by contributing additional ATP to phosphorylate carbon cycle intermediates and LHCP. This may set in motion the autocatalysis responsible for the attainment of high rates of photosynthetic carbon dioxide assimilation (Walker, 1981).

Further connections between the "light" and "dark" reactions are now becoming apparent, besides the supply of ATP and NADPH. The light driven exchange of Mg^{2+} and H^+ between the thylakoid inner space and stroma has been shown to have a strong influence on the activity of certain key enzymes, notably fructose *bis*phosphatase and sedoheptulose *bis*phosphatase (Heldt, 1979; Buchanan, 1980; Leegood and Walker, 1982).

It is not immediately obvious how this increased knowledge of the light reactions might be manipulated to further improve productivity. In theory the process of electron capture is fundamental to productivity, but in practice it is the subsequent utilization of these captured photons, after their conversion to energy rich compounds, which determines productivity. To some extent therefore, the amount of energy plants invest in the synthesis of the photochemical components of photosynthesis will be determined by the maximum demand for energy and reducing power in the dark reactions and for photosynthate for growth. Any improvement therefore would relate only to improving the efficiency of the light reactions.

Detailed analyses suggest that the potential storage efficiency of gross photosynthesis lies between 9.5 and 12% of solar energy (UK-ISES, 1976; Bassham, 1977; Bolton, 1978; Good and Bell, 1980; Beadle and Long, 1985). As this is already close to the practical maximum upper limit for a mechanism of light energy conversion there would appear to be little room for improving this efficiency (Bolton, 1978). This is supported from measurements of quantum yield ϕ under light-limiting conditions. In spite of marked differences in ϕ between C_3 and C_4 plants, there appears to be no difference between quantum yield within each group when comparisons are made under similar conditions (Ehleringer and Björkman, 1977).

At the same time, little information is available on the direct relationships between "light reactions" and productivity. Obviously, there can be no productivity without photochemistry, but in the field situation how frequently do "light reactions" limit the rate of dry matter accumulation? Although individual leaves of most C_3 plants are unable to utilize additional light above about 500 µmol m^{-2} s^{-1}, roughly ¼ full sunlight, this is not true of a crop canopy where shading ensures a continued increase in dry matter accumulation with increase in amount of light. Thus light is usually limiting to the productivity of an established crop. However, it is not clear whether the limitation is due to the photochemical production of NADPH + ATP or perhaps to the indirect control of stromal enzymes of carbon assimilation through the light reactions.

8.2 CARBON METABOLISM

The proliferation of studies on photosynthetic carbon metabolism since 1950 has not altered the basic view of the RPP cycle as proposed by Calvin and co-workers and its pivotal role in photosynthetic mechanisms. They have provided however a greatly improved knowledge of mechanisms of control of this unique autocatalytic cycle and its interactions with other functions of the green cell and plant as a whole (Latzko and Kelly, 1979; Robinson and Walker, 1981). When a leaf or isolated chloroplasts are illuminated after a period in the dark the increase in the rate of CO_2 assimilation (F) shows a lag of 30 s–3 min followed by a rise of 3–30 min before the maximum F and maximum efficiency of light energy conversion is reached (Walker and Robinson, 1978). In isolated chloroplasts the early phase of this lag probably represents the time taken for the activation of some enzymes of photosynthetic carbon metabolism through the events in the "light reactions". In particular, fructose 1:6-*bis*phosphatase (FbPase) and sedoheptulose 1:7-*bis*phosphatase (SbPase) appear to be strongly influenced by the reducing state of the stroma and by the light driven exchanges of protons and Mg^{2+} between the stroma and thylakoid inner space (Heldt, 1981; Buchanan, 1981; Halliwell, 1981). The subsequent rise in F_c following this lag represents the time taken for autocatalysis within the RPP-cycle to raise the level of intermediates, in particular the primary acceptor of CO_2 ribulose 1:5-*bis*phosphatase (Ru*b*P), to a maximum. Initially F increases exponentially with time following the lag, but will then plateau at a maximum which is probably determined by control mechanisms both inside and outside of the RPP-cycle (Latzko and Kelly, 1979). The maintenance of a steady-state assimilation of CO_2 must necessarily balance the regeneration of Ru*b*P from assimilated C against the loss of C to other metabolic pathways, a balance point which is reached when F reaches a maximum and remains constant. The pathway therefore requires internal feedback control and many possible mechanisms have been proposed (Bassham, 1979; Latzko and Kelly, 1979; Robinson and Walker, 1981; Cornic *et al.*, 1982). Competition for ATP by different reactions of the RPP cycle could also be important in its regulation. ATP is used to phosphorylate both PGA and Ru*b*P and if either component was in excess, phosphorylation of the other should be temporarily depressed. This has been suggested to account for secondary fluctuations in F following illumination of leaves and isolated chloroplasts (Robinson and Walker, 1981; Walker, 1981). These artificial dark-light transitions are of little direct relevance to the field situation. However, the speed with which a leaf can optimize its photosynthetic apparatus to a change in light levels, e.g. as a cloud crosses the sun, will clearly have a direct influence on production.

The amount of active enzyme as well as the amount of substrate must influence the maximum F. FbPase and SbPase which are considered to have an important regulatory role have activities more than adequate to account for measured rates of photosynthesis (Farquhar and von Caemmerer, 1982). The primary carboxylating enzyme Ru*bis*CO is often suggested to be in abundant supply, representing *ca.*

50% of the soluble protein of the chloroplast, at least in C_3 plants (Akazawa, 1977). However a number of studies have shown a good correlation between Ru*bis*CO activity and maximum F. A theoretical explanation of this is provided by the model of Farquhar *et al.* (1980).

In organisms as complex as higher plants it is to be expected that production of substances in photosynthesis will be closely linked to the organism's requirement for those substances. By way of illustration it is well known that artificial manipulation of the size of "sources" (e.g. the removal of leaves) or the size of "sinks" (e.g. the removal of storage organs) affects F, and that in general F increases with increase in the "sink" to "source" ratio. Further the bulk of C assimilated by the chloroplasts is exported from the leaf in the form of sucrose and in most leaves this export does not keep pace with the rate of photosynthetic C assimilation during daylight hours. This imbalance, manifested in the formation of starch in the chloroplasts, reduces F and is greater under conditions where translocation is inhibited, e.g. a reduced "sink" size, water stress, and temperature stress. Nevertheless, the supply of CO_2 can be shown to be partially limiting the rate of CO_2 assimilation in single leaves and crops under many conditions (Gaastra, 1959). At crop level, most important are high light and water stress when partial closure of stomata cause reductions in the CO_2 concentration inside the leaf. Even under cool temperate climatic conditions crop yields can be increased by increasing the ambient CO_2 concentration. Thus, photosynthetic C-metabolism, i.e. the "dark" reactions of photosynthesis, partially limit the rate of CO_2 uptake independent of the supply of NADPH and ATP from the "light" reactions. The "dark" limitation is supported by the observations given above that substrate levels, particularly Ru*b*P, may be only just sufficient under optimized conditions to support the maximum rates of CO_2 assimilation which are observed (Kelly *et al.*, 1975). Secondly, the "dark" reactions can limit the rate of CO_2 assimilation through photorespiration.

Photorespiration clearly accounts for loss of efficiency in the assimilation of CO_2 by C_3 species and these losses increase in relation to photosynthetic gains with increased temperature and water stress. Estimates of loss of assimilated C as a result of photorespiration, based on ^{14}C tracer studies range from 47% for tobacco (*Nicotiana tabacum*) to 17% for wheat (*Triticum aestivum*) (Zelitch, 1979). Even under cool temperate conditions it has been estimated that photorespiratory losses could amount to 20–60% of total dry matter production in *Triticum aestivum* (Keys *et al.*, 1977; Keys and Whittingham, 1981; Whittingham, 1981).

The absence of photorespiration in C_4 photosynthesis has two important consequences for production. First, since C_4 plants do not photorespire they can decrease their internal air space concentration of CO_2 almost to zero and thus maintain a greater inside–outside concentration gradient than that which C_3 plants can produce. This results in a greater amount of CO_2 being assimilated per unit amount of water transpired, i.e. a higher efficiency of water use. Secondly, individual leaves of most C_3 plants become light-saturated at photon flux densities well below full-sunlight because the lowered internal CO_2 concentration favours photorespiration. This explains why C_3 plants respond well to increases in the ambient CO_2 supply. In contrast, individual mature and healthy leaves of C_4 plants will not light saturate over the natural range of light levels and show higher efficiencies of light energy conversion at high light levels, i.e. 1 000 to 2 000 μmol m^{-2} s^{-1}. The greatest proportion of C_4 species in native floras are found in the hot semi-arid regions of the world where the gain from avoidance of photorespiration would be expected to be greatest.

In theory the productivity of C_3 crops could be increased by inhibiting photorespiration. One technique which has proved successful is to increase the ambient CO_2 concentration to a level where RubP oxygenation is inhibited; this is used now on a large scale in the greenhouse culture of *Lycopersicon esculentum* (tomatoes) where it can result in a 50% increase in yield (Warren-Wilson, 1972). Similar increases have been reported for *Triticum aestivum* grown in a high CO_2 concentration (Whittingham, 1981). However, this technique is not practicable at present on a field scale, since it requires enclosure of the atmosphere around the crop. Chemical inhibition of glycolate metabolism has also been shown to inhibit photorespiration and increase F in *Nicotiana tabacum* leaves in short term experiments (Zelitch, 1979) but could only be successful if glycolate metabolism is not essential to the functioning of the green cell. It has been suggested from a theoretical viewpoint that the relative affinities of Ru*bis*CO for CO_2 and O_2 could be altered (Ogren, 1978). However, since years of selection in natural environments where photorespiration is prominent have failed to do this in C_3 plants, the practical prospects of this seem poor at present.

CHAPTER 9
Gas-Exchange

9.1 CONTROLLED ENVIRONMENTS

9.1.1 The approach

As yield and photosynthesis are more often than not limited by weather, it is necessary to identify those factors which are most limiting for a particular crop or environment (Kramer, 1980). The measurement of photosynthesis in the field in response to natural complexes of environmental factors does not provide much information on the mechanism of the responses involved (Jarvis, 1970). Controlled environments have therefore been used to measure the response of single leaves (or plants of known leaf area) to an external or internal factor while all other factors are held constant. A comprehensive manual of techniques for measuring photosynthetic production is available (Sestak et al., 1971). Coombs and Hall (1982) also provide an introduction to this subject.

The development of a resistance analogue for gaseous diffusion into and out of plants has enabled photosynthesis to be expressed as the product of (i) a driving force, (ii) the CO_2 concentration gradient and (iii) a resistance (Gaastra, 1959). Two major resistances were obtained from gas-exchange analysis of photosynthesis (Chapter 1): the stomatal resistance, r_s between the leaf surface and the intercellular air space and the mesophyll resistance, r_m between the intercellular space and the site of fixation of CO_2 in the cell. The derivation and use of resistance analogues is explained by Jarvis (1971); the expression of mesophyll resistance on a cell wall area basis and the further partitioning of mesophyll resistance are considered by Nobel (1974, 1977, 1980b) and Prioul and Chartier (1977), respectively. Stomatal and mesophyll resistances of plants under "optimum" conditions are quite variable (Jarvis, 1971). The ratio between the two will determine the relative importance of each as factors limiting photosynthesis. In general $r_m : r_s$ is smaller in C_4 than in C_3 plants, owing to the more efficient fixation of CO_2 by C_4 plants. In gas-exchange terms therefore, much higher rates of photosynthesis are maintained in C_4 plants for the same CO_2 gradient between the intercellular space and the chloroplasts than in C_3 plants. A comprehensive list of stomatal resistances and rates of photosynthesis for a range of species is given by Körner et al. (1979).

Gas-exchange experiments in controlled environments have been mainly used to measure the effects of environmental variables on the photosynthesis of uniform plant material. This section lists a few of these variables and summarizes the type of information yielded from these studies. The plant material used in these experiments is often grown in controlled-environment growth cabinets or a glasshouse where partial control of the environment is possible.

9.1.2 Light

The responses of photosynthesis to light have been studied in cuvettes of many designs (e.g. Sestak *et al.*, 1971) but it is only in an integrating sphere, which surrounds the shoot with diffuse light of high photon flux density, that the true photosynthetic performance at a given photon flux density and the true photosynthetic capacities of plants at saturating photon flux density may be measured (Zewlawski *et al.*, 1973; Szaniawski and Wierzbicki, 1978). Apparent quantum yield has been measured in white light at very low flux densities, since accurate measurements of the maximum quantum yield (ϕ) can only be made on the initial linear part of the response curve of photosynthesis to light (Mohanty and Boyer, 1976; Ehleringer and Björkman, 1977; Ludlow, 1980; Monson *et al.*, 1982). The results of such experiments were reported in Chapters 4 and 6. More detailed analyses of absorption and action spectra, and relative quantum yields measured in light of different wavelengths in the visible spectrum have been reported elsewhere (McCree, 1972, 1981; Overdieck, 1979). The factors affecting the light compensation point (Ashton and Turner, 1979) and the effects of light pretreatment (Ludlow and Wilson, 1971b; Ludwig *et al.*, 1975) have also been studied. Further analysis and understanding of the light response curve with respect to leaf position (Rook and Corson, 1978), direction of illumination (Leverenz and Jarvis, 1979), shoot and stem photosynthesis (Begg and Jarvis, 1968), hysteresis effects with increasing and decreasing light (Ng and Jarvis, 1980), cuticular and stomatal phases (Ogawa, 1975) and photoinhibition (Osmond, 1981) have also been made through gas-exchange experiments.

The many gas-exchange experiments carried out to study the differences between sun and shade plants and adaptation to sun and shade have been reviewed by Boardman (1977) and Björkman (1981). Such experiments have been used to elucidate the ecological adaptation of ecotypes within a species to their respective environments (Eagles and Treharne, 1978).

9.1.3 Temperature

The temperature response curves of photosynthesis vary between species and between C_4 and C_3 plants (Murata and Iyama, 1963; Cooper and Tainton, 1968; Ludlow and Wilson, 1971a; Long *et al.*, 1975). Much of this information has been obtained from gas-exchange studies and reviewed by Berry and Björkman (1980).

Special precautions should be taken during the determination of photosynthesis and temperature response curves to maintain a constant water vapour pressure deficit (VPD) within the cuvette. The exponential increase of saturation vapour pressure with temperature may otherwise cause a decrease in photosynthesis at high temperature because of stomatal closure in response to high vapour pressure deficits, rather than through any direct effect of temperature on the photosynthetic process itself.

Controlled environments and cuvettes have provided a means for the study of adaptation to temperature (Downton and Slatyer, 1972; Doley and Yates, 1976; Bird et al., 1977; Mooney et al., 1977; Pearcy, 1977; Williams and Kemp, 1978), the seasonal variation of temperature optimum (Neilson et al., 1972; Slatyer, 1977a,b), and the relationship between photosynthesis and photorespiration (Lloyd and Woolhouse, 1976; Berry and Raison, 1981).

The interactions between light and temperature (Chabot and Chabot, 1977; Ku and Hunt, 1977; Long and Woolhouse, 1978) are of particular importance in relation to stress (Taylor and Rowley, 1971; Bagnall, 1979; Öquist et al., 1980; Martin et al., 1981; Long et al., 1983) and the effects of heat stress (Bauer, 1972) have all been measured by infra-red gas analysis.

9.1.4 Carbon dioxide and oxygen

Carbon dioxide is not considered as an important environmental variable except under unusual atmospheric conditions (Monteith and Sziecz, 1960; Lemon, 1963). It can be a very useful tool in gas-exchange studies, however, in the understanding of the limiting effects of other environmental variables on photosynthetic processes e.g. light (Bierhuizen and Slatyer, 1964; Brun and Cooper, 1967; McPherson and Slatyer, 1973; Ludwig et al., 1975; Wong et al., 1978; Wong, 1979b) and temperature (Troughton and Slatyer, 1969), and to distinguish between C_4 and C_3 plants (Chartier et al., 1970; Long and Woolhouse, 1978). The relationship between photosynthesis and intercellular CO_2 concentration has been related to the activity of Ru*bis*CO and the availability of Ru*b*P under CO_2 limiting and saturating conditions, respectively (Farquhar and Sharkey, 1982). The measurement of steady-state photosynthesis under varying ratios of $O_2 : CO_2$ has contributed to a proper understanding of the relationship between photosynthesis and photorespiration (Hesketh and Baker, 1967; Downes and Hesketh, 1968; Moore, 1977; Somerville and Ogren, 1979; Canvin et al., 1980) particularly in relation to temperature (Rowley and Taylor, 1972; Ku and Edwards, 1978; Peisker et al., 1979). Gas-exchange analysis can also be used to screen for differing sensitivity to CO_2 and O_2 within a species (Lloyd and Canvin, 1977; Powles et al., 1980).

9.1.5 Water status

The relationship between photosynthesis and plant water status has been studied with respect to water potential (e.g. Brix, 1962) or relative water content (e.g. Troughton, 1969). Stress treatments have been imposed on plants in several ways:
(a) withholding water from the growth medium (Boyer, 1970b; Beadle *et al.*, 1973; Mederski *et al.*, 1975; Ludlow and Ng, 1976; Bunce, 1977).
(b) differential levels of irrigation (Ackerson *et al.*, 1977).
(c) addition of polyethylene glycol (Gross, 1976; Lawlor, 1976a,b).
(d) by artificially restricting water uptake (Redshaw and Meidner, 1972) or
(e) excising shoots in air after steady-state rates of photosynthesis have been obtained (Beadle *et al.*, 1981).

Relative water content is measured after the experiment from leaf discs (Redshaw and Meidner, 1972) or directly during the experiment from a calibrated β-gauge using β-particles (Mederski *et al.*, 1975). Leaf-water potential is measured by a thermocouple psychrometer (Boyer, 1970a) or pressure bomb (Beadle *et al.*, 1981) at the end of the experiment or with a thermocouple dewpoint hygrometer placed *in situ* on the leaf just outside the chamber during the experiment (Beadle *et al.*, 1973). The major contribution from these types of experiments has been a better understanding of the regulation of photosynthesis by stomatal control and by mesophyll processes (Boyer, 1970b; Mederski *et al.*, 1975; Bunce, 1977; Beadle *et al.*, 1981; Monson *et al.*, 1982).

Stress treatments can also include those caused by atmospheric deficits and gas-exchange studies have focussed on the relationship between photosynthesis or stomatal conductance and VPD. The major effect of this stress appears to be stomatal closure with increasing VPD, though precautions should be taken to avoid extrapolating results obtained for single leaves to whole plants (Rawson *et al.*, 1977).

9.1.6 Other studies

The techniques of gas-exchange analysis have been applied to the study of windspeed (Caldwell, 1977a; Grace and Russell, 1977; Russell and Grace, 1978) and ontogenetic changes (Ludlow and Wilson, 1971c; Catsky *et al.*, 1976; Aslam *et al.*, 1977; Feller and Erismann, 1978). They have also been used as a starting point for steady-state labelling of leaves for the measurement of photorespiration (Ludwig and Canvin, 1971), for pulse and pulse-chase studies of photosynthetic metabolic pools (Mahon *et al.*, 1974; Ishii *et al.*, 1977; Lawlor *et al.*, 1977; Smith *et al.*, 1982), and to study feedback effects on photosynthesis on growth (Gifford, 1977).

Response curves obtained from plants grown in controlled environments are limited since the influence of environment and previous history, which affect performance in the field, cannot be adequately simulated in controlled conditions (Jarvis, 1970). It is also difficult to establish how representative of field behaviour the results from controlled environments really are (Biscoe *et al.*, 1975a). The precise reasons for these differences are not clear but in many cases are related to the environmental conditions experienced in growth cabinets *viz*. too low a photon flux density, unnatural spectral composition, and light received in such a way during growth as to be atypical of natural conditions (e.g. square wave for 16 hours).

The requirement for comparable plant material from glasshouses or controlled environments for gas-exchange studies has received less attention than its importance merits. The experiments of Warrington and Mitchell (1975, 1976), Warrington *et al.* (1976, 1978a,b) and Ludlow and Ng (1976) however, make it abundantly clear that proper attention to light quality and quantity are essential if plants similar to those growing under field conditions are to be obtained.

9.2 NATURAL ENVIRONMENTS

9.2.1 The approaches

Productivity represents the difference between the total influx and total efflux of CO_2 across a crop or plant community. Continuous monitoring of this exchange will provide an estimate of plant production though it may not be possible to distinguish that proportion which is harvestable without supplementary measurements e.g. by growth analysis and dry-weight determinations and considerations of the carbon content of the plant material. Productivity measurements by gas-exchange are also frequently difficult in natural communities with diverse species. Estimates of biomass production by gas analysis therefore have been attempted only in a few instances. The measurement of photosynthesis and carbon dioxide uptake in the field can be approached by several techniques. One technique measures the CO_2 flux over a uniform stand or crop. The second makes use of large chambers which enclose a group of plants, and the third is an extension of the cuvette method considered in the previous section which involves measuring leaves at all levels in the crop canopy. A fourth technique uses labelled carbon dioxide ($^{14}CO_2$) to measure photosynthetic rates of individual shoots. The first, and over a limited area the second, measure the external carbon budgets of the stand where the others seek to identify the distribution in space and time of photosynthetic activity within the canopy. The interpretation of these measurements in terms of discrete effects of single environmental factors on photosynthetic performance is problematical as many environmental variables change simultaneously and are partially or wholly related e.g. radiation and temperature, temperature and crop-atmosphere water vapour pressure deficit. To some extent, the results of experiments from controlled environments have successfully unravelled these interactions. Conversely, the experimental determination of photosynthesis in the field must be used to confirm results predicted from controlled environments as well as measuring field performance and the integrated plant response to habitat.

9.2.2 CO_2 flux

Three techniques are available for the measurement of CO_2 flux. Each have been used above crop canopies and are based on (a) momentum transfer or (b) energy budgets which use wind profiles and the Bowen ratio respectively, and (c) eddy correlation. These methods are considered by Denmead and McIlroy (1971), Thom (1975) and Ohtaki and Matsui (1982).

Examples of their use appear in case studies for several crops (see Montieth, 1976). Patterns of CO_2 flux are closely correlated with solar radiation (Denmead, 1969; Brown and Rosenberg, 1971; Eckardt et al., 1971). Micrometeorological techniques have also been used to measure CO_2 flux within plant canopies (Inoue et al., 1968; Saugier, 1970; Denmead, 1976; Ripley and Redmann, 1976). These studies have demonstrated that the major contribution to crop photosynthesis is from the upper leaves. In sparse canopies, the flux was found to be proportional to the ratio of leaf area duration to leaf area index (Uchijima, 1976). The contribution of the flux of CO_2 from the soil beneath the crop varies and is often ignored (Monteith, 1962; Lemon, 1967). It is not expected to increase the rate of photosynthesis by more than 2–4% (Brown, 1976).

9.2.3 Enclosure techniques

The complete enclosure of a group of plants within a large cuvette has been used to measure the photosynthetic rate of crop canopies. In contrast to measurements on single leaves in crops, enclosures avoid the necessity of sampling at different levels in the canopy, but do not identify sinks for carbon dioxide uptake. Like all assimilation chambers, several limitations are imposed by the creation of an artificial environment. These are mitigated by air-conditioning systems, but the eddy structure of wind close to the leaf surface and the radiation load will still differ from those outside.

Enclosure techniques have been used to measure the seasonal dynamics of carbon dioxide exchange in grass swards and salt marsh vegetation (Sheehy, 1977; Redmann, 1978; Drake and Read, 1981). Seasonal changes of photosynthesis in a mixed grassland in Winnipeg at the same time of day (1400 h) were similar throughout a three-year period, increasing between April and July and then decreasing as high temperature accentuated the effects of water-stress (Ripley and Redmann, 1976). Puckridge (1969) used the same technique to estimate the photosynthetic contribution of different plant parts by removing them sequentially.

9.2.4 Cuvettes

The most popular technique for measuring photosynthesis and plant production has used monitoring of single shoots enclosed in cuvettes at several points in the canopy. An artificial environment is still created around the shoot but the precise distribution of sinks for carbon dioxide within the canopy can be identified.

A review of some of the early work with cuvettes was given by Schulze and Koch (1969). Ventilated chambers with Peltier-controlled air-conditioning systems (Koch et al., 1968) have been used to study field photosynthesis of plants in response to temperature under desert conditions (Lange et al., 1974, 1975, 1978) and the performance of Norway spruce particularly in relation to the significance of the evergreen habitat (Fuchs et al., 1977; Schulze et al., 1977a,b). The techniques and limitations of these experiments have been discussed by Schulze et al. (1972). Similar measurements with air conditioned chambers have been made on Scots pine (Troeng and Linder, 1982a,b). A list of further studies using cuvettes in coniferous canopies appeared in Jarvis et al. (1976). A gas-exchange technique developed by Littleton (1971) was used to measure crop photosynthesis in a stand of barley (Biscoe et al., 1975c). Light response curves for each organ were measured using natural variations in irradiance or imposing variation, if necessary, using a white paper sleeve as a neutral density filter. Crop photosynthesis for any one hour was then calculated from the mean distribution of irradiance with height and the summed contribution of each organ using the corresponding light response curve. Parkinson and Day (1983) have varied CO_2 as well as irradiance in a study of drought effects on the growth of *Hordeum sativum* in the field.

9.2.5 Labelled carbon dioxide

An alternative technique for estimating the photosynthesis of single leaves measures the uptake of $^{14}CO_2$ (Austin and Longden, 1967; Incoll and Wright, 1969; Shimshi, 1969; Turner and Incoll, 1971; Coombs and Hall, 1982; Bell and Incoll, 1982). The total uptake of carbon dioxide is then calculated from the specific activity of the gas. The method has been further developed by Incoll (1977) and for coniferous species by Neilson (1977). Temperature-controlled exposure chambers adaptable to all leaf types have also been described (Bingham and Coyne, 1977).

Exposure times normally vary between 10s and 60s and, unlike cuvettes in systems incorporating laboratory infra-red gas analysers, enable extensive sampling of plant canopies within short periods of time (Landsberg et al., 1975; Watts et al., 1976). There was close agreement between this technique and a cuvette method in the responses of photosynthetic rate to light in *Hordeum sativum* and when the two methods were used simultaneously on leaves of the C_4 species *Spartina anglica* (Biscoe et al., 1977; Long and Incoll, 1979).

9.2.6 Comparisons

A few experiments have compared carbon budgets or plant production estimated from two or more techniques. The most comprehensive set of data suggested close agreement could be obtained between (i) dry weights estimated from CO_2 fluxes and growth analysis (Biscoe *et al.*, 1975b), and (ii) crop photosynthesis from CO_2 fluxes and a cuvette technique (Biscoe *et al.*, 1975c). Similar comparisons have been made for grasslands between an enclosure chamber and a micrometeorological technique (Ripley and Redmann, 1976) and between an enclosure chamber and growth analysis (Sheehy, 1977). Takeda *et al.* (1976) and Wielgolaski; (1977) compared cuvette and harvesting techniques with good agreement on rice and alpine communities, respectively.

CHAPTER 10

Limits to Biomass Production

10.1 LIGHT INTERCEPTION AND GROWTH

The canopy photosynthesis of crops and natural communities will be determined by the product of their average photosynthetic rate and leaf area index. The gross productivity of the community will then equal the integrated value of canopy photosynthesis throughout the growing season. As there is a linear relationship between canopy photosynthesis and light interception, this is a major factor controlling productivity and the length of the growing season is a major determinant of the maximum quantity of light which may be intercepted (Fig. 10.1; Monteith, 1981).

During the early part of the growing season, after germination of crop plants, after defoliation of forage crops by cutting or grazing, and in young plantations of trees, interception is closely related to the available leaf area since radiation falling on the soil will not be absorbed by the photosynthetic apparatus (Loomis and Williams, 1969; Rhodes, 1973). Early in crop growth, dry matter production is proportional to the rate of absorption of light (Monteith, 1965b, 1981) which is causally related to the rapid increase in canopy photosynthesis as a result of increasing leaf area index until canopy closure occurs (Ludwig et al., 1965; Hayashi, 1966; King and Evans, 1967; Wilson and Teare, 1972). Measured relationships between dry matter production and leaf area index early in the season are found to be linear (Ashley et al., 1965; Eik and Hanway, 1966; Iwata and Okubo, 1971); this is also true for dry matter production or canopy photosynthesis and the percentage interception of solar radiation (Williams et al., 1965; Shibles and Weber, 1966; Baker and Meyer, 1966; Hesketh and Baker, 1967; Gallagher and Biscoe, 1978). In this way the total interception of light by crop canopies can often be an accurate predictor of total photosynthesis and biomass production. After canopy closure, the relationship between interception and available leaf area is no longer linear and the distribution of radiation through the canopy becomes more important. The effects of mutual shading of leaves does however ensure a continued increase in dry matter production with increase in the amount of light. It is in this way that light is usually limiting to the productivity of an established crop, though the amount intercepted cannot of course exceed that incident above the crop. In practice the relationships between crop growth rate (C), or canopy photosynthesis, and the

Fig. 10.1 *The relationship between total dry matter production of four crops grown in the United Kingdom with intercepted (total) solar radiation, illustrating the importance of light interception as a major determinant of plant production and canopy photosynthesis (reproduced with permission from Monteith, 1977).*

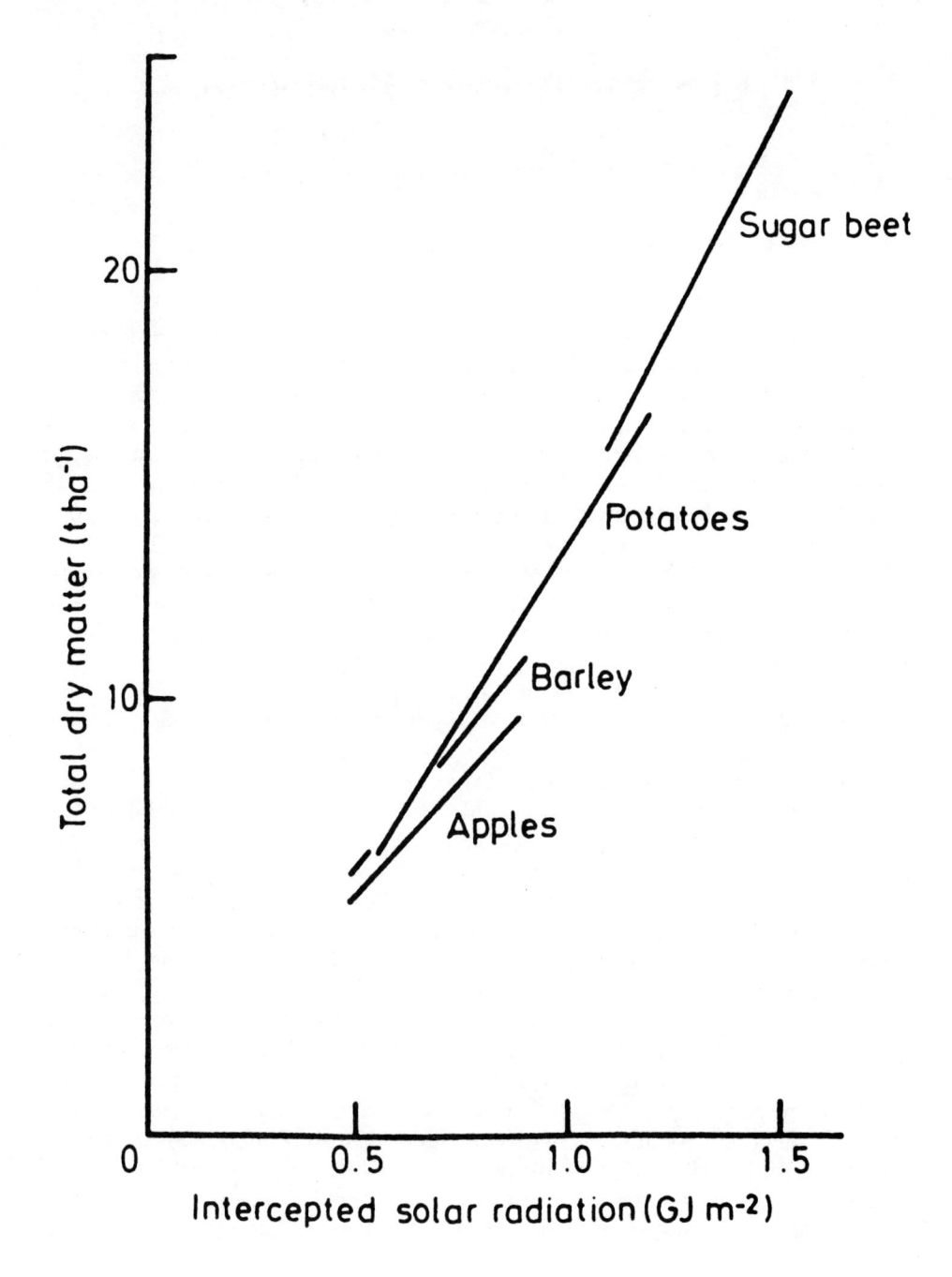

available leaf area or leaf area index (L), differs with the crop concerned and this may reflect the growth habit of the plant. Some species are observed to have an optimum L (Davidson and Donald, 1958; Stern and Donald, 1962; Black, 1963; Harper, 1963) while others show an asymptotic relationship for C against L, where beyond a critical L, C remains constant (Brougham, 1958; Loomis, 1963; Shibles and Weber, 1965; Williams et al., 1965; McCree and Troughton, 1966; Wifong et al., 1967). King and Evans (1967) suggested that in prostrate species, less light would penetrate to the base of the canopy and a greater proportion of actively respiring tissue would be shaded and act as a sink for assimilates, thus producing an optimum L. If leaves adapt physiologically to a shade environment however, an optimum L should not be observed (Loomis and Williams, 1969) since respiration is not directly proportional to leaf area but is adjusted to be a constant fraction of the growth rate (McCree and Troughton, 1966). Maximum crop growth rate in wheat was associated with a critical L though there was still a tendency for C to fall at very high indices (Fischer et al., 1976). This subject is reviewed further by Yoshida (1972).

10.2 CANOPY STRUCTURE

From the many species examined to date, the photosynthetic rate per unit leaf area of single leaves does not appear to be an important factor governing productivity. Nevertheless productivity will ultimately depend on the total photosynthetic performance of the canopy and the photosynthetic capacity of the individual shoots. Self-shading between leaves reduces both performance and capacity of individual leaves since each develops and photosynthesizes in a restricted light environment. Considerable attention has therefore been given to manipulating the structure of crops to enhance the light interception and photosynthetic efficiency of the leaf area available. An alternative would be to breed for specific light absorption and transmission characteristics (Woolhouse, 1978) but the optical properties of leaves show little interspecific variation (Kleshin and Shuglin, 1959; McCree, 1972; Gausmann et al., 1973). It is therefore necessary to study characteristics which relate to foliage posture to determine how plant morphology can be manipulated to increase productivity.

Leaf inclination was first suggested as an important factor for dry matter production by Boysen-Jensen (1932, 1949). For many erect-leaved (erectophile) or horizontal-leaved (planophile) species, the characteristic pattern of leaf arrangement may seem qualitatively obvious but species can be categorized more exactly with respect to their leaf arrangement. For example, foliage distributions are considered to fall into three categories: uniform, random, or clumped (see Loomis et al., 1971). Alternatively, the variance : mean ratio obtained from the variance of foliage contacts using a simple point quadrat (Warren-Wilson, 1965) may be used to determine the light penetration through the canopy. The variance equals the mean if the distribution is random. Ratios less than one indicate a regular and those greater than one a contagious distribution. Other methods of measuring leaf inclination are reviewed by Trenbath and Angus (1975). Leaf angle is not a constant property of crops and may change systematically and irreversibly with age as well as being correlated with leaf size (Ledent, 1978; Drake and Turitzin, 1980). Erect leaves are also generally shorter than lax (or droopy) leaves (Tanaka et al., 1966; Angus et al., 1972; Blad and Baker, 1972). A further parameter of plant canopies can be described by Beer's Law (Monsi and Saeki, 1953; Kasanaga and Monsi, 1954). This equation describes attenuation of light as exponential with respect to the leaf area index of the canopy and assumes that this does not change with depth (Ludwig et al., 1965; Newton and Blackman, 1970). In general, the extinction coefficient is low for erect (i.e. more light penetration) and high for horizontal leaves, but this will vary with solar angle and also cloudiness (Kuroiwa and Monsi, 1963; Anderson, 1966).

The quantification of the effects of leaf inclination on canopy photosynthetic rates and yield is difficult. The common practice is to use mathematical descriptions of model canopies to predict the effects of changes of area, position, and angle of leaves on the interception of light (Anderson, 1964; Duncan, 1971; Bonhomme and Chartier, 1972; Ross, 1975). This approach prompted Monteith (1969) to suggest that there was a "curious reluctance" to test models of light penetration with measurements of rates of photosynthesis and dry matter production in the field. In spite of the limitations of these types of models (see Baker et al., 1977 for review) comprehensive studies in real canopies may however require a formidable number of measurements. In general, if L is less than 2, these models predict that a crop with prostrate leaves is expected to have a higher gross photosynthetic rate than crops with an erect-leaf habit. Should L be greater than between 4–8 then the reverse is true (Monteith, 1969; Trenbath and Angus, 1975). An alternative approach has made use of photographic techniques to derive the probability of canopy elements intercepting light (Evans and Coombe, 1959; Fuchs, 1972; Fuchs and Stanhill, 1980). This is accurate for light interception regardless of spatial distribution and appears to agree with radiometric measurements on different crops (Stanhill et al., 1972; Fuchs and Stanhill, 1980).

The results of several field trials which compare the productivity of cultivars of differing leaf inclination are listed by Trenbath and Angus (1975). The advantage in percentage terms of the more erect-leaved cultivar of any one species is always positive for crop growth rate but more variable for grain yield except for rice where the advantage of erect leaves appear considerable and has contributed to the successful development of the crop (Owen, 1968; Chandler, 1969; Tanaka et al., 1969; Akiyama and Yingchol, 1972).

The most suitable method for distinguishing the effect of inclination from other factors is to use isogenic material differing only in leaf erectness. As these are only available for maize (Trenbath and Angus, 1975), it becomes more difficult to establish causal relationships between canopy yield and morphology for most crops. Yield advantages for maize hybrids with upright leaves were only small (0.4

to 5.6%) when compared with hybrids of normal canopy structure at their respective optimum L (Winter and Ohlrogge, 1973). It would appear that positive yield responses to leaf erectness are difficult to identify in maize except possibly at high L and high planting densities (Pepper et al., 1977; Gerakis and Papakosta-Taspoula, 1980) though these advantages may be spurious if heavy self-shading causes a high percentage of barren plants in lax-leaved hybrids (Pendleton et al., 1968; Trenbath and Angus, 1975).

The advantage of erect-leaved cultivars should be more noticeable in the C_3 cereals which are grown at higher densities, have higher L, and lower saturating quantum flux densities for photosynthesis than maize, a C_4 species (Evans and Wardlaw, 1976; Monteith, 1977). Some evidence suggests that erect leaves in barley, wheat and oats may be correlated with yield (Tanner et al., 1966) but no differences were detected between two wheat cultivars of contrasting canopy structure (Puckridge and Ratkowsky, 1971). In a further experiment an erectophile cultivar of the same species had consistently higher levels of canopy photosynthesis throughout the season, but this was counteracted by differences in translocation of stored materials in stems (of the lax-leaved cultivars) which resulted in similar grain yields (Austin et al., 1976).

It might be expected that the advantages of erect tissue would be expressed to a greater extent in crops where the economic yield constitutes a high proportion of the dry matter, for example forage grasses under frequent cutting management. To some extent this is the case, but the matter is not clear-cut. The crop growth and canopy photosynthetic rates of three temperate forage grasses of contrasting structure were essentially identical (Sheehy, 1977; Sheehy and Peacock, 1977). Rhodes (1973) has suggested that leaf rigidity, number of tillers per plant, leaf size and tiller angle may be more important determinants of canopy structure and yield in herbage grasses than leaf angle which shows little variation within species. In this case, morphological characters other than leaf angle may be predominantly associated with yield (Jones et al., 1979).

The relationship between leaf orientation and yield in crops is complex. Canopy morphology can be influenced by agronomic practice including planting density and row direction. Some species even exert heliotropic responses (e.g. Fukai and Loomis, 1976). Although breeding for crop canopies which correspond to an ideotype (Donald, 1968) may lead to some yield improvement, in practice it appears that differences in leaf angle are of lesser significance to the final yield of many crops than the rate at which the canopy expands to form a complete cover of the ground (Monteith, 1965b; Fischer et al., 1976). The relationships between leaf inclination and crop production are reviewed at length by Rhodes (1973) and Trenbath and Angus (1975).

10.3 GROWTH ANALYSIS

A major contribution to the understanding of bioproductivity, particularly of crop plants, has come through the application of plant growth analysis. Growth analysis may be distinguished from direct measurements of photosynthesis which give much information about fundamental physiological process and its short-term response to the environment. In contrast, growth analysis has provided quantitative measurements of actual biomass production over longer intervals of days or weeks which may also be related to short term variations in the environment (Hunt, 1978). Further, growth analysis provides a means for identifying the relative importances of changes in leaf area and net assimilation per unit leaf area, in explaining variation in production between and within species. The techniques applied to growth analysis have their own inherent limitations (Monteith, 1981) but they allow evaluation of the importance of photosynthetic efficiency in its contribution to yield.

The historical development of growth analysis has been reviewed by Watson (1952) and Baker *et al.* (1977) and the application of growth analysis has been described by Radford (1967) and Hunt (1978). The relative growth rate, R describes the rate of production of dry weight per unit of initial dry weight in terms of a compound interest law (Blackman, 1919). It can be expressed instantaneously (R) or as an average value over a preselected period ($\bar{R}$) and measures the efficiency of the plant as a producer of new material (Watson, 1952). There are considerable differences in R within species which relate to marked interactions of ontogeny and environment (Duncan and Hesketh, 1968; Eagles, 1969; Grime and Hunt, 1975; Roberts and Wareing, 1975). The value of R does not identify the underlying reasons for these differences.

For further analysis, relative growth rate (R) may be expressed as a product of the unit leaf rate (E) and the leaf area ratio (A). E and A measure the efficiency of the plant or crop as a producer of dry weight and leaf area respectively. The product of unit leaf rate (E) and the leaf area index (L) measures the instantaneous crop growth rate (C) which serves as a simple index of agricultural productivity. The corresponding values over a discrete time period are $\bar{E}$, $\bar{A}$, $\bar{C}$, and $\bar{L}$ (Hunt, 1978; Beadle, 1982). As E and $\bar{E}$ measure the rate of dry matter production per unit of leaf material, it may be viewed as a direct measurement of photosynthetic rate less respiratory losses on the same basis, and is thus a measure of assimilatory efficiency. E varies in a complex manner with environmental factors, but increases linearly with light in crop canopies; it is also highly correlated to seasonal changes in the light environment, at least in temperate environments (Watson, 1947). This correlation may break down after full crop cover as changes in E are masked by changes in L causing mutual shading of leaves. There is also a wide variation in E between species which may be linked to the type of photosynthetic mechanism used (Coombe, 1960; Jarvis and Jarvis, 1964; Monteith, 1978; Dunn, 1981) but except in specific instances, L is a more important determinant of C than is E (Watson, 1952). The leaf area index (L) is a measure of the leaf area of the crop per unit land area but does not assume any particular arrangement of the leaves. Unlike E, L is greatly affected by seasonal variations in environmental stress and by competition from other species and weeds in mixed and arable crop communities, respectively. L is also a function of the phenology of any one particular species and the time of planting for annual crops (Watson, 1947; Watson, 1971). A more useful expression for L is a measure of its persistence. This is the integrated area under the curve relating L with time and is referred to as leaf area duration, D. The product of E and D is a measure of weight per unit area or yield, though it is inevitably a crude measure as E is unlikely to remain constant throughout the season and D will contain components with differing photosynthetic efficiency (Hunt, 1978). The relative importance of variation of L and D in crops has been discussed at length by Watson (1952) who concludes that in general they are more important than E as a determinant of crop yield.

The use of conventional growth analysis (Watson, 1947) was limited by the lack of information during the intervals between harvests and failed to account for changes in the parameters of growth analysis with time (ontogenetic changes). This was particularly crucial in the determination of E where a linear relationship between leaf area and dry weight was assumed, and large errors resulted if it was strongly non-linear (Radford, 1967). The use of curve fitting and regression procedures has transformed the practice of growth analysis into a series of small harvests over short intervals, and the utilization of all the available harvests in determining values at any point in time. Appropriate functions which adequately describe the appearance of the data are then fitted (Richards, 1959; Vernon and Allison, 1963; Hughes and Freeman, 1967; Nicholls and Calder, 1973; Causton et al., 1978; Venus and Causton, 1979). The development of computer programs for fitting these functions has made the use of this approach more practicable (Causton, 1969; Hunt and Parsons, 1974; Hunt and Parsons, 1977). The advantages and disadvantages of the functions are discussed elsewhere (Radford, 1967; Hurd, 1977; Hunt, 1978).

The relationships between photosynthetic rate and the components of growth analysis are complex. In general, high photosynthetic rates were directly proportional to the concentration of water soluble carbohydrates in the early stages of growth in tall fescue and E and R were greater for high than for low yielding varieties but in general genotypes were high yielding regardless of photosynthetic rate (Wilhelm and Nelson, 1979). Correlations between rate of net photosynthesis measured over short periods and biomass yield are poor (Heichel and Musgrave, 1969; Evans and Dunstone, 1970; Charles-Edwards, 1971; Rhodes, 1972). No single factor appears to explain this poor correlation, though leaf area production, measured in terms of L or D, appears to be the more important influence on yield (Hanson, 1971; Yoshida, 1972; Hoveland et al., 1974). For example, strong correlations between L and C were observed at all stages of growth in *Zea mays* and maximum L was correlated with seed yield in *Brassica* (Clark and Simpson, 1978). The storage and use of carbohydrates for the rapid production of leaf area, particularly at the start of the season, and the distribution of dry matter or assimilate partitioning throughout the period of crop growth are thought to be the basic factors which influence the relationship between net photosynthetic rate and yield and which determine yield itself (Treharne and Eagles, 1970; Yoshida, 1972; Potter and Jones, 1977; Good and Bell, 1980). For example, the partitioning of a greater proportion of newly assimilated dry matter into leaf tissue during the first 16–19 days of regrowth was highly correlated with yield in *Festuca arundinacea* (Wilhelm and Nelson, 1979). Seed yield was correlated with assimilate partitioning in soybean rather than the toal production of dry matter (Shibles and Weber, 1966).

10.4 IN THE CONTEXT OF OTHER FACTORS LIMITING PHOTOSYNTHETIC EFFICIENCY

It is perhaps surprising that correlations between the rate of net photosynthesis per unit leaf area and biomass production are generally poor (Heichel and Musgrave, 1969; Evans and Dunstone, 1970). This apparent contradiction arises because of the negative correlation between net photosynthetic rate and leaf or mesophyll cell size (Tsunoda, 1962; Wilson and Cooper, 1967; Evans and Dunstone, 1970) and because photosynthetic efficiency is constrained both directly and indirectly by other limiting factors. As these indirect limiting factors are of immediate relevance to the improvement of photosynthetic productivity, they will be briefly considered.

A major reason for this poor correlation, briefly considered in the last section, relates to the leaf area of the crop. The product of photosynthesis and leaf area determines the total production of dry matter and not the individual leaf photosynthetic rate *per se*. Furthermore, the total area of the leaf surface, an integration of the rate of expansion, leaf area duration, and senescence of leaves will directly influence total productivity.

Another determinant of yield is the partitioning and use of photosynthate, and it appears that the interaction between photosynthesis and its use or storage could be a major factor limiting yield. Thus photosynthate or dry matter may be used for the production of economic yield or specific storage organs and will be required for respiration in order to synthesize new and to maintain existing tissues. Alternatively, photosynthate may be used for the production of further photosynthetic tissues, i.e. new leaves. Past emphasis on increasing crop yields has been through enhancement of plant "sink" and "storage" capacity, and increasing the harveset index i.e. the proportion of the plant which is harvested. The rate of photosynthesis of most cultivated crops is less than that of their progenitors (Nasyrov, 1960; Evans, 1975; McArthur *et al.*, 1975; Vergara, 1977). Their photosynthetic potential per unit leaf area does not therefore appear to have been a factor which determined the role that "source" size played in overall performance and productivity of cultivated species, but more the distribution of dry matter between storage and the production of new photosynthetic area. The creation of new leaf results in both new "sinks" for the products of photosynthesis and maintains or increases the rate of photosynthesis *per se* (Good and Bell, 1980). The factors which govern the phenology i.e. leaf area and distribution of the crop are therefore of crucial importance to the expression of its inherent photosynthetic potential and biomass yield. The rate of translocation from the leaf, and the precise nature of the feedback controls which limit photosynthesis when translocation is restricted from the leaf, are also of importance here, in addition to the question of whether the total size of the "source", the "sink" or the two in combination limit yield (Neales and Incoll, 1968; Evans, 1975; Gifford and Evans, 1981).

There were few attempts by plant breeders in the past to improve the efficiency of the photosynthetic apparatus, though new techniques in gene technology may now make this more feasible (Day, 1977), particularly as the genetic base for plastid inheritance becomes better understood (Nasyrov, 1978). Some attempt has been made to improve the efficiency of species with low photosynthetic rates. This has been approached by hybridization of *Atriplex* species possessing the C_3 and C_4 pathways of photosynthesis. However the photosynthetic rates of both F_1 and F_2 hybrids was less than that of both parents (Björkman *et al.*, 1971; Nobs, 1976). The genetic constraints associated with the Kranz leaf anatomy apparently make it impossible to impart the properties of compartmentation of photosynthesis, essential to C_4 photosynthesis, into C_3 plants (Nasyrov, 1978). Experimental mutagenesis has also been used to seek increased photosynthetic efficiency. In a few instances, positive results have been obtained (Highkin *et al.*, 1969; Usamanov *et al.*, 1975) but the considerable shifts in biochemical and structural organisation which normally occurs during mutagenesis has usually given negative results. The association of C_4 plants with high PEP carboxylase activity and the considerable variation of PEP carboxylase activity which occurs in C_3 plants, suggests that the photosynthetic efficiency of C_3 plants might be enhanced by screening for this enzyme. There is also evidence that the kinetic rate constants of Ru*bis*CO may vary from plant to plant (Seeman and Berry, 1982). Attempts to screen for low photorespiration indicate that there are varieties within populations of C_3 species with this characteristic, but it is not yet clear whether it is associated with greater productivity (Wilson, 1972; Zelitch, 1973; Zelitch and Day, 1973). Further, the absence of good techniques for accurate measurements of photorespiration cast doubt on the validity of these studies.

Brown *et al.* (1976) have identified other key factors which limit the photosynthetic productivity of crops. These have been considered at length in recent reviews (Wittwer, 1977; Wittwer, 1980a,b) and include non-photosynthetic genetic improvement, efficiency of nutrient uptake, biological nitrogen fixation, resistance to competing biological systems (pests and weeds), resistance to environmental stresses, and hormonal mechanisms in relation to plant development.

CHAPTER 11

Integrative Models relating Photosynthesis to Productivity

Part III has, so far, considered aspects of photosynthetic research from the cellular to the whole plant level. In contrast, classical plant physiology uses reductionist techniques to isolate sublevel processes from each other and from those at a higher level. As we have attempted to use an integrative approach to identify the complexity of factors which limit biomass production at a photosynthetic level, mathematical models have been increasingly employed as one way to effect integrations between photosynthesis, crop growth and the complexities of the environment. The primary aim of constructing these models has been to simulate photosynthetic performance or biomass production of plants over a wide range of conditions. When completed and validated, using data which has been collected independently from that used to construct the model, responses to changes in the environment or to new environments for the plant, which may take years to establish experimentally, may then be simulated in moments.

The general approaches to modelling in crop physiology including photosynthesis and crop growth have been reviewed extensively in recent years (Baker *et al.*, 1977; Loomis *et al.*, 1979; Hesketh and Jones, 1980; Singh *et al.*, 1980; Charles-Edwards, 1982). The complexity of these models has varied considerably but tend to include detailed considerations of environmental factors and canopy geometry in conjunction with more simple descriptions of photosynthesis at the biochemical level. The potential usefulness of these simulation models has not always been realised but their primary aim remains as a tool to increase the understanding of processes which control productivity.

Thornley (1976) has examined a number of equations which describe relationships between single leaf photosynthesis and some environmental variables. These have been incorporated into models which describe the effects of light, carbon dioxide and oxygen on leaf photosynthesis (Charles-Edwards and Ludwig, 1974; Chartier and Prioul, 1976; Tenhunen *et al.*, 1977) and some models include dark respiration (Hall and Björkman, 1975). The purpose of these models has been to identify functional relationships which include realistic parameters for characterizing the photosynthetic process, for example quantum efficiency and light saturated photosynthetic rate of a light response curve.

Mechanistic models provide a means by which the integrated biophysical and biochemical steps of photosynthesis at the sub-cellular level may be related to leaf, and ultimately crop photosynthesis. Thus, the implications for leaf photosynthesis of known effects of environmental factors on individual sub-cellular processes and hypotheses on the importance of different sub-cellular processes in limiting photosynthesis may be tested. This has already been illustrated in the models of Farquhar et al. (1980) and Farquhar and von Caemmerer (1982) which have clearly illustrated the limitations imposed by availability of Ru*bis*CO or Ru*b*P under different conditions of CO_2 supply. Such modelling will provide a link between the hitherto separate areas of sub-cellular photosynthesis research on the one-hand and gas-exchange and production research on the other, and will allow examination rather than speculation on the relationship between knowledge in the two areas.

Simpler mechanistic models of this type have been developed to cope with the response of photosynthesis in single leaves to the fluctuating environment in the field (Reed et al., 1976). Known responses of net photosynthesis to important environmental variables are predetermined in laboratory experiments or by imposing environmental treatments in the field (Biscoe et al., 1975b) and applied to data collected in the field where many environmental variables are correlated with one another. The major advantages of this approach are the production of parameters which characterize the photosynthetic capacity of the foliage and a means of predicting the relative impact of each environmental variable on photosynthesis (Reed et al., 1976; Detling et al., 1978). In contrast, models based on multiple linear regression techniques or which do not account for all environmental variables (e.g. Chartier, 1969, 1970; Hall, 1971; Lommen et al., 1971; Taylor and Sexton, 1972) contain parameters which have no immediate identity.

A recognition that it is the area of foliage and its light interception which determine crop growth, rather than the photosynthetic efficiency of the leaves, has led to the development of sophisticated light distribution models as the basis for modelling of canopy photosynthesis. Most of these models have assumed a random or horizontal distribution of leaf area, although other distributions are probably important, particularly in row crops (Allen and Brown, 1965; Ross and Nilson, 1967; Acock et al., 1969). Several models have failed to account for the absorption of light by plant parts other than leaves and have assumed that the interception of light by plant canopies remains constant throughout the day although it is clear that the extinction coefficient can vary with sun and leaf angle (Isabe, 1962; Monteith, 1965a; de Wit, 1965; Hesketh and Baker, 1967). The Duncan model (Duncan et al., 1967) for simulating photosynthesis in crop canopies included corrections for solar elevation and like de Wit's model (1965) utilized different light response curves for different canopy layers. The model confirmed the advantages of an erect-leaved canopy at high leaf area index and a horizontal canopy at low leaf area index. Further developments have allowed for the variation of leaf angle and light response curve through the canopy and have resulted in more complex models (Verhagen et al., 1963; Kuroiwa, 1969; Ross, 1969; Ross, 1975; Goudriaan, 1977).

SIMCOTT II (see Baker et al., 1977) and COTTON (Stapleton et al., 1973) are models which measure light interception based on (i) leaf area index and (ii) on plant height and row width, respectively. This second approach was found to be more effective for row crops where the height rather than density of the foliage is of more importance for light interception. Models for canopy photosynthesis have also been included in micrometeorological models which simulate energy transport, transpiration and CO_2 exchange in plant canopies (Stewart and Lemon, 1969; Waggoner, 1969; Murphy and Knoerr, 1970, 1972; Sinclair et al., 1971). Less complex models have since been developed which identify the more important constraints on photosynthesis, such as light within plant canopies, and assume others to deviate little from ambient conditions through the canopy (Sinclair et al., 1976). Predictions of CO_2 assimilation were within 12% of the complete model, and a third model which assumed that all the leaves were exposed to the same microenvironment also agreed closely with the full model.

In many crop models where yield prediction is the primary objective photosynthesis has been included as the major process in the ecosystem which determines productivity. Estimates of gross photosynthesis from such models, after correction for respiratory losses, are considered to be good predictors of net primary production (de Wit, 1965; Alberda and Sibma, 1968). Respiratory losses have also been the subject of modelling techniques (Penning de Vries et al., 1974) and these have been incorporated subsequently into crop growth models (Hunt and Loomis, 1979).

The major models relating photosynthesis to crop productivity are either static, i.e. involve no concept of time, or dynamic. The latter employ a hierarchic approach and can be used to provide prediction and explanation from a knowledge of morphological, physiological and biochemical processes. In general, static models (e.g. Thompson, 1969; Murata, 1975; Bridge, 1976; Nelson and Dale, 1978; Pitter, 1977) employ a multivariate regression approach and after some manipulation are the most suitable means for yield prediction (Loomis et al., 1979). Dynamic models have been based on simulating photosynthetic productivity in environments which under optimal conditions vary chiefly with radiation. The development of dynamic models have been considered by Loomis et al (1979) and those with more detailed hierarchic structures consist of very large numbers of variables e.g. BACROS (de Wit, 1978) and SUBGOL (Fick et al., 1973, 1975; Hunt and Loomis, 1979). Simpler models based on the above have also been developed following sensitivity analyses which identify the relative importance of each factor.

Further development of dynamic models has led to the prediction of economic yields. These models use partitioning factors and can be used to demonstrate the potential yield improvements that might be possible from changing aspects of partitioning (Duncan et al., 1978) as well as being of general use as predictors of food production under various agricultural strategies (Buringh and Heemst, 1977). The demonstration that yield increases in the short-term are more likely to result from changes in partitioning rather than improvements in the photosynthetic capacity of single leaves (de Vries et al., 1967; Evans, 1975; Good and Bell, 1980) suggest that this type of model deserves further development.

CHAPTER 12

Future Perspectives

The majority of the world's people live by growing plants or processing their products, and thus depend on the productivity of plants for their wellbeing. Biomass production in terrestrial environments must therefore continue to rise as long as it remains necessary to meet the demands of a growing world population. Increased food production is an essential part of overall biomass production in a world where 600 million people are estimated to-day to be seriously undernourished and hungry. For the world's rural poor who constitute nearly three-quarters of the world's population, biomass production is also their main source of fuel, clothing fibres and building materials. Uneven distribution of fossil energy resources and their increasing costs indicate that competition between rich and poor for existing and potentially new plant resources will inevitably intensify. Can the world achieve, and the environment sustain, the increased photosynthetic production of plant materials that it will need in increasing quantity?

Many of the requirements of present plant production systems and our mistreatment of the environment would argue that this may not be possible. Much of the growth in food crop production in the 60s and 70s was achieved by the increased use of non-renewable and energy intensive resources in the form of fertilizers, pesticides and mechanization. New cereal varieties able to respond strongly to high rates of fertilizer application were developed and, in some instances, tripled yields. Secondly, an increase in the world's cultivated area by deforestation proceeded at a rate of 11 million hectares per annum supplying new agricultural land. Much marginal land was made productive by irrigation. There is evidence that the environment cannot continue to sustain this increase in crop production which could be negated in both very obvious and also in subtle ways. The poor compete for the energy intensive non-renewable resource of inorganic fertilizers because high yielding and disease resistant varieties of the major cereals often require high fertilizer inputs. Deforestation, especially in Amazonia, has already influenced patterns of rainfall and soil water content adversely, and may possibly also have longer term effects on atmospheric CO_2 levels.

There is an increasing realization that a soundly-based and well supported plant production "industry" (agriculture and forestry) is essential for the wellbeing of individual countries or regions. Industrialization which had been hoped to free the dependence of the economies of many developing countries from a dependence of plant products may often be causing serious environmental problems for plant production (e.g. Agarwal *et al.*, 1982). Atmospheric pollutants emanating from industrial processes and motor exhausts, in the developed countries, are known to significantly reduce crop production through their direct effect on photosynthesis. This is in spite of the introduction of emission controls. Effects on natural temperate vegetation are still poorly understood and the information is patchy, but industrialization of developing countries is probably producing similar environmental problems. It is unfortunate that these countries may too often be unable to afford safeguards against pollution or may be tempted to lure investment by imposing less rigorous emission controls.

In W. Europe and N. America atmospheric pollution has significantly damaged and sometimes eliminated major component species of natural ecosystems. Significant damage often occurs some distance, up to 500 km, from the source of pollution. Thus, it is not simply vegetation close to industrial centres that is damaged; indeed it may not even be the same country which suffers. If effects following industrial development seen in the older industrialized countries are repeated in tropical countries large tracts of natural vegetation will be lost. Indeed, effects on the already precarious existence of vegetation in the semi-arid tropics may well be more serious and could conceivably enhance desertification at an unpredictable rate and with unpredictable effects on atmospheric CO_2 and ground water levels. The gloomy prospects for the maintenance of current levels of production does not remove these problems and much effort will be needed in the future to meet the goals of self-sufficiency in food and other plant products. What is the best approach?

Apart from improved fertilization of the land, improvements in pest resistance, pest protection and harvest index have been chiefly responsible for crop yield increases in the last two decades. Harvest index is the proportion of total crop biomass represented by that portion which is harvested. However, there is a limit to yield improvement by these techniques and in the major cereals, where total biomass production has remained constant (Austin, 1980), the point may be being approached where more and more effort wil be required for ever decreasing gains. However, the potential limit to crop production is set by crop photosynthesis and yet little attention has been paid to the possibility of increasing total biomass production through photosynthesis. The only notable exception has been the I.R. rice varieties which were specifically selected for a better canopy architecture and light distribution between leaves so raising crop photosynthesis. Even this improvement is governed by the dependence of many of these varieties on high fertilizer inputs.

Improvement of crop photosynthesis therefore may be the necessary step, in the longer term, to meet the rising and presently unmet world demand for plant products. As the practical maximum efficiency of photosynthetic conversion of solar energy into energy trapped in plant matter is about 5–6% and that achieved by crops rarely exceeds 2%, with averages below 1%, there would seem to be much room for improvement. To assess if this improvement is possible and what effects such improvement would have on the global environment, especially atmospheric CO_2, a complete world picture is needed. Whilst we have a reasonably detailed fundamental knowledge of photosynthesis of the most important crops of temperate countries and some knowledge of photosynthesis in natural vegetation, the equivalent knowledge for the greater diversity of crops and natural ecosystems of developing countries, i.e. two-thirds of the land mass, is lacking.

Having accepted that increased photosynthesis is the key to increased biomass production, it must be appreciated that the primary factor which determines the rate of dry matter production is leaf canopy photosynthesis. The importance of canopy size, its speed of formation and its duration has been illustrated by the close correlation between intercepted radiation and crop production for a number of temperate crops. A good understanding of both leaf development and photosynthetic production processes would appear essential for substantial yield improvements. Photosynthetic studies at the cellular level have so far contributed little to improving bioproductivity but since cellular and molecular processes ultimately determine photosynthetic efficiency, improved knowledge of their role as limiting factors is essential, particularly so if developments in recombinant-DNA technology are to be fully exploited. An unknown quantity, but clearly fundamental to photosynthesis, is the increasing CO_2 concentration in the atmosphere. This will affect productivity at the ecological, whole plant and cellular level via carbon fixation and efficiency of water use. Further studies are therefore urgently required into this form of pollution as the consequences of elevated CO_2 concentration are as yet poorly understood.

The foregoing chapters have considered the mechanisms which relate to all stages of the photosynthetic process where photosynthesis could potentially limit production. The relative importance of these mechanisms however cannot at present be established since the diffuse nature of the contents found in the literature make valid comparisons difficult. While there are abundant studies of all phases of the photosynthetic process, it is rare to find any two studies from different laboratories which use the same species and cultivars grown under similar environmental conditions. The same criticism also applies to field studies of photosynthesis and productivity from different parts of the world, since these seem rarely to use directly comparable methods. One hope for the future would be greater co-operation between research workers in the use of a few standard cultivars. This must not, of course, preclude investigation of photosynthesis under sub-optimal conditions i.e. where environmental factors such as temperature, salinity, and water stress, impose severe constraints on photosynthetic performance. The current paucity of studies, particularly at the cellular level in these areas prevents an understanding of the effects of these stresses on limitations within the photosynthetic apparatus.

The further study of plant canopy architecture and leaf area development as major components for increasing light interception and hence photosynthetic productivity in crop stands, should be encouraged. Similarly, research into leaf area development and partitioning of photosynthate between leaves and other plant parts, particularly in relation to the production of leaf area prior to full canopy closure in agricultural crops and forests, will lead to improvements in crop photosynthesis. Other limitations, *viz.* leaf area development, resource partitioning, and genetic constraints at the level of the photosynthetic apparatus contribute to the determination of overall bioproductivity (Brown *et al.*, 1976). It is essential nevertheless to exploit the physiology and biochemistry of plants as a tool for increasing production – since it is crop photosynthesis which ultimately sets the upper limit on improved bioproductivity.

REFERENCES

Ackerson, R. C., Krieg, D. R., Haring, C. L. & Chang, N. (1977) *Crop Sci. 17*, 81.
Acock, B., Thornley, J. H. M. & Wilson, J. (1969) In: *Prediction and measurement of photosynthetic productivity* (I. Setlik, ed.) p. 91. Wageningen.
Agarwal, A., Chopra, R. & Sharma, K. (1982) *The state of India's environment.* Centre for Science and Environment, New Delihi.
Akazawa, T. (1977) In: *Proc. 4th Int. Phot. Cong.* (D. O. Hall, J. Coombs, T. W. Goodwin, eds.) p. 447. The Biochemical Society, London.
Akiyama, T. & Takeda, T. (1975) *Proc. Crop Sci. Soc. Japan. 44*, 6.
Akiyama, T. & Yingchol, P. (1972) *Proc. Crop Sci. Soc. Japan. 41*, 126.
Alberda, T. & Sibma, L. (1968) *J. Br. Grassl. Soc. 23*, 206.
Alberte, R. S., Thornber, J. P. & Fiscus, E. L. (1977) *Plant Physiol. 59*, 351.
Allaby, M. (1977) *World food resources* 418 pp. Applied Science
Allen, L. H. & Brown K. W. (1965) *Agron. J. 57*, 575.
Anderson, J. (1982) *Photochem. Photobiophys. 3*, 225.
Anderson, J. & Melis, A. (1983) *Proc. Natl. Acad. Sci. 80*, 745.
Anderson, J. W., & Done, J. (1977a) *Plant Physiol. 60*, 354.
Anderson, J. W., & Done, J. (1977b) *Plant Physiol. 60*, 504.
Anderson, K., Shanmugam, K. T., & Valentine, R. C. (1977) In: *Genetic engineering for nitrogen fixation* (A. Hollaender, ed.) p. 95. Plenum Press.
Anderson, L. E. (1975) In: *Proc. 3rd Int. Phot. Cong.* (M. Avron, ed.) p. 67. Elsevier.
Anderson, M. C. & Denmead, O. T. (1969) *Agron. J. 61*, 867.
Anderson, M. C. (1964) *J. Ecol. 52*, 27.
Anderson, M. C. (1966) *J. Appl. Ecol. 3*, 41.
Angus, J. F., Jones, R. & Wilson, J. H. (1972) *Aust. J. Agric. Res. 23*, 845.
Anon (1978) *Plant damage caused by SO_2.* Environmental directorate. Report Workshop 7-8th June, 1978. OECD, Paris.
Anon (1982) *Emissions, Costs and Engineering Assessment.* Work Group 3B Report No. 38. Final. US-Canada MOI on Transboundary Air Pollution.
Apel, P. (1980) *Biochem. Physiol. Pfl. 175*, 386.
Arkcoll, D. B. & Festenstein, G. N. (1971) *J. Sci. Food Agric. 32*, 49.
Armond, P. A., Schreiber, U. & Björkman, O. (1978) *Plant Physiol. 61*, 411.
Arnon, D. I. (1977) In: *Encycl. Plant Physiol.* N.S. Vol. 5 (Avron, M., Trebst, A. A. Eds.) p. 7. Springer-Verlag.
Arthur, M. A. (1982) In: *Climate in Earth* History. (W. H. Berger, J. C. Crowell, eds.) p. 55 National Academy Press, Washington D.C.
Ashley, D. A., Doss, B. D. & Bennet, O. L. (1965) *Agron. J. 57*, 61.
Ashton, D. J. & Turner, J. S. (1979) *Aust. J. Bot. 27*, 589.
Aslam, M., Lowe, S. B. & Hunt, L. A. (1977) *Can. J. Bot. 55*, 2288.
Austenfeld, F. A. (1979) *Photosynthetica 13*, 434.
Austin, R. B. (1980) In: *Opportunities for increasing crop yields.* (R. G. Hurd, P. V. Biscoe, C. Dennis, eds.) p. 3. Pitman.
Austin, R. B., Ford, M. A., Edrich, J. A. & Hooper, B. E. (1976) *Ann. Appl. Biol. 83*, 425.
Austin, R. B. & Longden, P. C. (1967) *Ann. Bot. 31*, 245.
Bacastow, R. B. (1981) In: *Global Energy Futures and the Carbon Dioxide Problem* (G. Speth, e. d.) p. 84 Council of Environmental Quality. Washington D.C.
Badger, M. R. & Collatz, G. J. (1977) *Carn. Inst. Wash. Yearbk. 76*, 355.

Baes, C. F., Goeller, M. E., Olsen, J. S. & Rotty, R. M. (1976) In: *The global carbon dioxide problem.* Report ORNL-51 94, p. 1. Oak Ridge National Laboratory, Tennessee.
Bagnall, D. (1979) In: *Low temperature stress in crop plants: the role of the membrane* (J. M. Lyons, D. Graham, J. K. Raison, eds.) p. 67. Academic Press.
Baker, D. N., Hesketh, J. D. & Weaver, R. E. C. (1977) In: *Crop Physiology* (V. S. Gupta, ed.) p. 110. Oxford & IBH Publ., New Delhi.
Baker, D. N. & Meyer, R. E. (1966) *Crop Sci. 61,* 15.
Baker, N. R. (1978) *Plant Physiol. 62,* 889.
Baker, N. R. (1984) In: *Topics in Photosynthesis Vol. 5* (J. Barber, N. R. Baker, eds.), Elsevier. In press.
Baker, N. R., East, T. M. & Long, S. P. (1983) *J. Exp. Bot. 34,* 189.
Ballschmitter, K. & Katz, J. J. (1972) *Biochem. Biophys. Acta. 256,* 307.
Barber, J. (1977) *The intact chloroplast. Vol. 2.* 516 pp. Elsevier.
Barber, J. (1982) *Ann. Rev. Plant Physiol. 33,* 261.
Barber, J. (1983) *Plant Cell Env. 6,* 311.
Barghoorn E. S. (1984) In: *Paleobotany, Paleoecology and Evolution* (J. Stone, ed.), Praeger. In Press.
Barr, T. N. (1981) *Science, 214,* 1087.
Bassham, J. A. (1977) *Science, 197,* 630.
Bassham, J. A. (1979) In: *Encycl. Plant Physiol. N.S. Vol. 6* (M. Gibbs, E. Latzko, eds.) p. 9. Springer-Verlag.
Bassham, J. A., & Buchanan, R. B. (1983) In: *Photosynthesis* Vol. 2 (Govindjee ed.) p. 141. Academic Press.
Bauer, M. (1972) *Photosynthetica 6,* 424.
Bauer, M. (1979) *Z. Pflanzenphysiol. 92,* 277.
Bauer, M. & Senser, M. (1979) *Z. Pflanzenphysiol. 91,* 359.
Bazzaz, F. A. (1980) *Abstr. 5th Int. Phot. Cong.* p. 45.
Bazzaz, F. A., Rolfe, G. L. & Carlson, R. W. (1974a) *Physiol. Plant. 32,* 373.
Bazzaz, F. A., Rolfe, G. L. & Windle, P. (1974b) *J. Env. Qual. 3,* 156.
Beadle, C. L. (1982) In: *Techniques in bioproductivity and photosynthesis* (J. Coombs, D. O. Hall, eds.) p. 20. Pergamon Press.
Beadle, C. L. & Jarvis, P. G. (1977) *Physiol. Plant. 41,* 7.
Beadle, C. L., Jarvis, P. G., Neilson, R. E. & Talbot, H. (1981) *Physiol. Plant. 52,* 391.
Beadle, C. L. & Long, S. P. (1985) *Biomass,* In press.
Beadle, C. L., Stevenson, K. R., Neumann, H. H., Thurtell, G. W. & King, K. M. (1973) *Can. J. PLant Sci. 53,* 537.
Beckerson, D. W. & Hofstra, G. (1979a) *Atmos. Env. 13,* 1263.
Beckerson, D. W. & Hofstra, G. (1979b) *Can. J. Bot. 57,* 1940.
Begg, J. E. & Jarvis, P. G. (1968) *Agric. Meteorol. 5,* 91.
Bell, C. J. & Incoll, L. D. (1982) *J. Exp. Bot. 33,* 896.
Bennett, J., Steinbeck, K. E., Arntzen, C. J. (1980) *Proc. Natl. Acad. Sci. 77,* 5253.
Bennett, J. H. & Hill, A. C. (1973) *J. Air Poll. Cont. Ass. 23,* 203.
Bennett, K. J. & Rook, D. A. (1978) *Aust. J. Plant Physiol. 5,* 231.
Bernstein, L. & Hayward, H. E. (1958) *Ann. Rev. Plant Physiol. 9,* 25.
Berry, J. A. (1975) *Science 138,* 644.
Berry, J. A. & Björkman, O. (1980) *Ann. Rev. Plant Physiol. 31,* 491.
Berry, J. A., Fork, D. C. & Garrison, S. (1975) *Carn. Inst. Wash. Yearbk. 74,* 751.
Berry, J. A. & Raison, J. K. (1981) In: *Encycl. Plant Physiol. N.S. Vol. 12A* (O. L. Lange, P. S. Nobel, C. B. Osmond, H. Ziegler, eds.) p. 278. Springer-Verlag.
Bethlenfalvay, G. J., Abu-Shakra, S. S. & Phillips, D. A. (1978a) *Plant Physiol. 62,* 127.
Bethlenfalvay, G. J., Abu-Shakra, S. S. & Phillips, D. A. (1978b) *Plant Phyusiol. 62,* 131.
Bidinger, F., Musgrave, R. B., & Fischer, R. A. (1977) *Nature 270,* 431.
Bierhuizen, J. & Slatyer, R. O. (1964) *Aus. J. Biol. Sci. 17,* 348.
Biggs, R. H., Sisson, W. B., & Caldwell, M. M. (1975) In: *Impacts of climatic change on the*

REFERENCES

biosphere: Part 1 (D. S. Nachtway, M. M. Caldwell, R. H. Biggs, eds.) p. 4. U.S. Dept. Transport, Washington D.C.
Billings, W. D., Godfrey, P. J., Chabot, B. F. & Bourque, B. P. (1971) *Arct. Alp. Res. 3*, 277.
Bingham, G. E. & Coyne, P. I. (1977) *Photosynthetica 11*, 148.
Bird, I. F., Cornelius, M. J. & Keys, A. (1977) *J. Exp. Bot. 28*, 519.
Biscoe, P. V., Clark, J. A., Gregson, K., McGowan, M., Monteith, J. L. & Scott, R. K. (1975a) *J. Appl. Ecol. 12*, 227.
Biscoe, P. V., Scott, R. K. & Monteith, J. L. (1975b) *J. Appl. Ecol. 12*, 269.
Biscoe, P. V., Gallagher, I. N., Littleton, E. J., Monteith, J. L. & Scott, R. K. (1975c), *J. Appl. Ecol. 12*, 295.
Biscoe, P. V., Incoll, L. D., Littleton, E. J. & Ollerenshaw, J. H. (1977) *J. Appl. Ecol. 14*, 293.
Björkman, O. (1973) *Photophysiology Vol. 8*, (A. C. Giese ed.) p. 1. Academic Press.
Björkman, O. (1975) *Carn. Inst. Wash. Yearbk. 74*, 748.
Björkman, O. (1981) In: *Encycl. Plant Physiol. N.S. Vol. 12A* (O. L. Lange, P. S. Nobel, C. B. Osmond, H. Ziegler, eds.) p. 57, Springer-Verlag.
Björkman, O., & Badger, M. (1977) *Carn. Inst. Wash. Yearbk. 76*, 346.
Björkman, O., & Badger, M. (1979) *Carn. Inst. Wash. Yearbk. 78*, 262.
Björkman, O., Badger, M. & Armond, P. (1978) *Carn. Inst. Wash. Yearbk. 77*, 262.
Björkman, O., Badger, M. & Armond, P. (1980) In: *Adaptations of plants to water and high temperature stress* (N. C. Turner, P. J. Kramer, eds.) p. 231. Wiley-Interscience.
Björkman, O., Boardman, N. K. Anderson, J. M., Thorne, S. W., Goodchild, D. J., & Pyliotis, N. A. (1972) *Carn. Inst. Wash. Yearbk. 71*, 115.
Björkman, O., Boynton, J. & Berry, J. (1976) *Carn. Inst. Wash. Yearbk. 75*, 400.
Björkman, O., Mooney, H. A. & Ehleringer, J. (1975) *Carn. Inst. Wash. Yearbk. 74*, 743.
Björkman, O., Nobs, M. A., Pearcy, R., Boynton, J. E. & Berry, J. (1971) *Photosynthesis and Photorespiration* (M. D. Hatch, C. B. Osmond, R. O. Slatyer, eds.) p. 105. Wiley-Interscience.
Black, C. C. (1973) *Ann. Rev. Plant Physiol. 24*, 253.
Black, J. N. (1963) *Aust. J. Agric. Res. 14*, 20.
Black, V. J. (1982) In: *Effects of gaseous air pollution in agriculture and horticulture* (M. H. Unsworth, D. P. Ormrod, eds.) p. 67. Butterworths.
Black, V. J. & Unsworth, M. H. (1979a) *J. Exp. Bot. 30*, 81.
Black, V. J. & Unsworth, M. H. (1979b) *J. Exp. Bot. 30*, 473.
Black, V. H. & Unsworth, M. H. (1979c) *Nature 282*, 68.
Black, V. H. & Unsworth, M. H. (1980) *J. Exp. Bot. 31*, 667.
Blackman, V. H. (1919) *Ann. Bot. 23*, 353.
Blackwood, G. C. & Miflin, B. J. (1976) *J. Exp. Bot. 27*, 735.
Blad, B. L. and Baker, D. G. (1972) *Agron. J. 64*, 26.
Bledsoe, C. S. (1976) *Plant Physiol. 57*, Suppl. 49.
Boardman, N. K. (1977) *Ann. Rev. Plant Physiol. 28*, 355.
Boardman, N. K. (1978) In: *Proc. 4th Int. Cong. Phot.* (D. O. Hall, J. Coombs, T. W. Goodwin, eds.) p. 635. The Biochemical Society, London.
Boardman, N. K., Anderson, J. M., Thorne, S. W., & Björkman, O. (1972) *Carn. Inst. Wash. Yearbk. 71*, 107.
Böhning, R. H. & Burnside, C. A. (1956) *Amer. J. Bot. 43*, 557.
Bolhar-Nordenkampf, H. R. (1982) In: *Techniques in bioproductivity and photosynthesis* (J. Coombs, D. O. Hall, eds.) p. 58. Pergamon Press.
Bolin, B., (1977) *Science. 196*, 613.
Bolin, B., Degens, E. T., Duvigneaud, P., & Kempe, S. (1979) In: *Carbon Cycle Modelling*, Scope Report Number 13 (B. Bolin, ed.) p. i Wiley.
Bolin, B., Degens, E. T., Duvigneaud, P. & Kempe, S. (1979) *The global carbon cycle* (B. Bolin, E. T. Degens, S. Kempe & P. Ketner, eds.) p. 1. Wiley.

Bolin, B., ed. (1981) *Carbon Cycle Modelling.* Scope Report No. 16, 390 pp.
Bolton, J. K. & Brown, R. H. (1980) *Plant Physiol. 66,* 97.
Bolton, J. R. (1978) In: *Proc. 4th Int. Phot. Cong.* (D. O. Hall, J. Coombs, T. W. Goodwin, eds.) p. 621. The Biochemical Society, London.
Bonhomme, R. & Chartier, P. (1972) *Isr. J. Agric. Res. 22,* 53.
Bonte, J. (1975) PhD. Thesis, University of Pierre and Marie Curie, Paris.
Bortitz, S. (1964) *Biol. Zentralbl. 83,* 501.
Bottrill, D. E. Possingham, J. V. & Kriedemann, P. E. (1970) *Plant Soil, 32,* 424.
Bowen, G. D. (1980 In: *Contemporary microbial ecology* (D C. Ellwood, J. H. Hedger, M. J. Latham, J. M. Lynch, J. H. Slater, eds.) p. 283. Academic Press.
Boyer, J. S. (1970a) *Plant Physiol. 46,* 233.
Boyer, J. S. (1970b) *Plant Physiol. 46,* 236.
Boyer, J. S. (1971) *Plant Physiol. 48,* 532.
Boyer, J. S. (1976a) *Phil. Trans. R. Soc. Lond. B. 273,* 501.
Boyer, J. S. (1976b) In: *Water deficits and plant growth, Vol. IV* (T. T. Koslowski, ed.) p. 153. Academic Press.
Boyer, J. S. & Potter, J. R. (1973) *Plant Physiol. 51,* 989.
Boyer, J. S. & Youmis, H. M. (1983) In: *Effects of stress on photosynthesis* (R. Marcelle, H. Clijsters, M. van Poucke, eds.) p. 29. Junk.
Boyko, H. (1966) *Salinity and Aridity: New Approaches to Old Problems,* 408 pp. Junk.
Boysen-Jensen, P. (1932) *The Dry Matter Production of Plants.* 108 pp. Fischer-Verlag.
Boysen-Jensen, P. (1949) *Det. Kgl. Danske Videnskabernes Selskab, Biologiske Neddelelser 21.*
Bradford, K. J. & Hsaio, T. C. (1982) In: *Encycl. Plant. Physiol. N. S. Vol. 12B* (O. L. Lange, P. S. Nobel, C. B. Osmond, H. Ziegler, eds.) p. 263. Springer-Verlag.
Brandle, J. R., Campbell, W. F., Sisson, W. B. & Caldwell, M. M. (1977) *Plant Physiol. 60,* 165.
Bray, J. R. (1959) *Tellus. 11,* 220.
Bridge, D. W. (1976) *Agric. Meteorol. 17,* 185.
Briens, M. & Larher, F. (1982) *Plant Cell Env. 5,* 287.
Brix, H. (1962) *Physiol. Plant. 15,* 20.
Brougham, R. W. (1958) *Aust. J. Agric. Res. 9,* 39.
Brown, A. W. A., Byerly, T. C., Gibbs, M. & San Pietro, A. (eds.) (1976) *Crop productivity – research imperatives.* 399 pp. Mich. Agric. Expt. Sta.
Brown, H. & Escombe, F. (1905) *Proc. Roy. Soc. Lond. B. 76,* 118.
Brown, K. W. (1976) In: *Vegetation and the atmosphere* Vol. 2 (J. L. Monteith, ed.) p. 65 Academic Press.
Brown, K. W. & Rosenberg, N. J. (1971) *Agron. J. 63,* 207.
Brown, L. R. (1975) *Science 190,* 1053.
Brown, L. R. (1981) *Science 214,* 995.
Brown, R. H. (1978) *Crop Sci. 18,* 93.
Brown, R. H. & Brown W. V. (1975) *Crop Sci. 15,* 681.
Brown, W. V. (1977) *Memoirs Torrey Bot. Club 23,* 1.
Brun, W. A. & Cooper, R. L. (1967) *Crop Sci. 7,* 451.
Buchanan, B. (1980) *Ann. Rev. Plant Physiol. 31,* 341.
Buchanan, B. (1981) In: *Proc. 5th Int. Phot. Cong. Vol. 4* (G. Akoyonoglou, ed.) p. 245. Balaban International, Philadelphia.
Bulen, W. A. & Le Compte, J. R. (1966) *Proc. Natl. Acad. Sci. 56,* 979.
Bull, J. N. & Mansfield, T. A. (1974) *Nature, 250,* 443.
Bull, T. A. (1969) *Crop Sci. 9,* 726.
Bunce, J. A. (1977) *Plant Physiol. 59,* 348.
Bunnik, N. J. J. (1978) *Meded. Landbhoogesch. Wageningen 78,* 1.
Buringh, P. (1980) *Limits to the productive capacity of the biosphere. Future sources of organic raw materials. Chemrawn I* (L. E. St. Pierre, G. R. Brown, eds.) p. 325.

Pergamon Press.
Buringh, P. & van Heemst, H. D. J. (1977) *An estimate of world food production based on labour-oriented agriculture* 46 pp. Wageningen.
Buringh, P., van Heemst, H. D. J. & Staring, G. J. (1975) *Computation of absolute maximum food production of the world.* 59 pp. Wageningen.
Burnside, C. A. & Böhning, R. M. (1957) *Plant Physiol. 32,* 61.
Butler, W. L. (1978) *Ann. Rev. Plant Physiol. 29,* 345.
Caldwell, M. M. (1971) *Photophysiology Vol. 4* (A. C. Giese, ed.,) p. 131. Academic Press.
Caldwell, M. M. (1977a) *Plant Physiol. 46,* 535.
Caldwell, M. M. (1977b) *Research in photobiology* (A. Castellani, ed.) p. 597. Plenum Press, New York.
Caldwell, M. M. (1981) In: *Encycl. Plant Physiol. N. S. Vol. 12A* (O. L. Lange, P. S. Nobel, C. B. Osmond, H. Ziegler, eds.) p. 169. Springer-Verlag.
Caldwell, M. M., Osmond, C. B. & Nott, D. (1977) *Plant Physiol. 60,* 157.
Campbell, W. H. & Black, C. C. (1978) *Bio System 10,* 253.
Canvin, D. T., Berry, J. A., Badger, M. K., Fock, H. & Osmond, C. B. (1980) *Plant Physiol. 66,* 302.
Capron, T. M. & Mansfield, T. A. (1976) *J. Exp. Bot. 27,* 1181.
Carlson, R. W. (1979) *Env. Polln. 18,* 159.
Carlson, R. W., Bazzaz, F. A., & Rolfe, G. (1975) *Env. Res. 10,* 113.
Catsky, J., Ticha, I. & Solarova, J. (1976) *Photosynthetica 10,* 394.
Causton, D. R. (1969) *Biometrics 25,* 401.
Causton, D. R., Elias, C. O. & Hadley, P. (1978) *Plant Cell Env. 1,* 163.
Chabot, B. F. & Chabot, J. F. (1977) *Oecologia 26,* 363.
Chandler, R. F. (1969) In: *Physiological aspects of crop yield.* (J. D. Eastin, F. A. Hoskins, C. Y. Sullivan, C. H. van Bavel, eds.) p. 265. American Society of Agronomy, Madison, Wisconsin.
Chang, C. W. & Heggestad, H. E. (1974) *Phytochemistry 13,* 871.
Chapman, S. B. (1976) *Methods in plant ecology.* 536 pp. Blackwell.
Charles-Edwards, D. A. (1971) *J. Exp. Bot. 22,* 663.
Charles-Edwards, D. A. (1982) *Physiological determinants of crop growth* 161 p. Academic Press.
Charles-Edwards, D. A. & Ludwig, L. J. (1974) *Ann. Bot. 38,* 921.
Chartier, P. (1969) *Ann. Physiol. Veg. 11,* 221.
Chartier, P. (1970) In: *Prediction and measurement of photosynthetic productivity* (I. Setlik, ed.) p. 307. Wageningen.
Chartier, P., Chartier, M. & Catsky, J. (1970) *Photosynthetica 4,* 48.
Chartier, P. & Prioul, J. L. (1976) *Photosynthetica 10,* 20.
Clark, J. M. & Simpson, G. M. (1978) *Can. J. Plant Sci. 58,* 587.
Cooke, G. W. (1975) *Fertilizer use and protein production,* p. 29. Potash Institute, Berne.
Coomb, D. E. (1960) *J. Ecol. 48,* 219.
Coombs, J. & Hall, D. O. (1982) *Techniques in bioproductivity and photosynthesis* 171 pp. Pergamon Press.
Coombs, J., Hall, D. O. & Chartier, P. (1983) *Plants as solar collectors: Optimizing productivity for energy.* 210 pp. D. Reidel Publ., Dordrecht.
Cooper, J. P. (1975) *Photosynthesis and productivity in different environments.* 715 pp. Cambridge University Press.
Cooper, J. P. & Tainton, N. M., (1968) *Herb. Abstr. 38,* 167.
Cornic, G., Ulso, K. C. & Osmond, C. B. (1982) *Plant Physiol. 70,* 1310.
Cote, W. A. (1983) *Biomass utilization.* Plenum Press.
Coulson, C. L. & Heath, R. L. (1974) *Plant Physiol. 53,* 32.
Cowling, D. W., Jones, L. H. P. & Lockyear, D. R. (1973) *Nature, 243,* 479.
Cowling, D. W. & Koziol, M. J. (1979) *J. Exp. Bot. 29,* 1029.
Coyne, P. I. & Bingham, G. E. (1978) *J. Air. Pollut. Cont. Assoc. 28,* 1119.

Critchley, C. (1982) *Nature 298*, 483.
Croughan, T. P. & Rains, D. W. (1982) In: *CRC Handbook of biosolar resources Vol. 1 Part 2* (A Mitsui, C. C. Black, eds.) p. 245. CRC Press, Boca Raton.
Dahlman, R. C. & Kucera, C. L. (1965) *Ecology 46*, 84.
Darbyshire, B. & Steer, B. T. (1973) *Aust. J. Biol. Sci. 26*, 591.
Davidson, J. L. & Donald, C. M. (1958) *Aust. J. Agric. Res. 9*, 52.
Day, W. (1981) In: *Physiological processes limiting plant productivity* (C. B. Johnson, ed.) p. 199. Butterworths.
Day, P. R. (1977) *Science 197*, 1334.
Denmead, O. T. (1969) *Agric. Met. 6*, 357.
Denmead, O. T. (1976) In: *Vegetation and the atmosphere. Vol. 2* (J. L. Monteith, ed.) p. 1. Academic Press.
Denmead, O. T. & McIlroy, I. C. (1971) In: *Plant Photosynthetic Production* (Z. Sestak, J. Catsky, P. G. Jarvis, eds.) p. 461., Junk.
Derwent, R. G. & Stewart, H. N. M. (1973) *Atmos. Env. 7*, 385.
Detling, J. K., Parton, W. J. & Hunt, H. W. (1978) *Oecologia 33*, 137.
Döbereiner, J., Burris, R. L. & Hollaender, A., eds. (1978) *Limitations and potentials for biological nitrogen fixation in the tropics.* Plenum Press.
Doley, D. & Yates, D. J. (1976) *Aust. J. Plant Physiol. 3*, 471.
Doliner, L. J. & Jolliffe, P. A. (1979) *Oecologia, 38*, 23.
Dominy, P. J. & Baker, N. R. (1980) *J. Exp. Bot. 31*, 59.
Donald, C. M. (1968) *Euphytica 17*, 385.
Downes, R. W. & Hesketh, J. D. (1968) *Planta 73*, 79.
Downton, J. & Slatyer, R. O. (1972) *Plant Physiol. 50*, 518.
Downton, W. J. S. (1975) *Photosynthetica 9*, 96.
Drake, B. G. & Raschke, K. (1974) *Plant Physiol. 53*, 808.
Drake, B. G. & Read, M. (1981) *J. Ecol. 69*, 405.
Drake, B. G. & Turitzin, S. N. (1980) *Abstr. 5th Int. Cong. Phot.* p. 151.
Duncan, W. G. (1971) *Crop Sci. 11*, 482.
Duncan, W. G., & Hesketh, J D. (1968) *Crop Sci. 8*, 670.
Duncan, W. G., Loomis, R. S., Williams, W. A. & Hanau, R. (1967) *Hilgardia 4*, 181.
Duncan, W. G., McCloud, D. W. McGraw, R. L. & Boote, K. T. (1978) *Crop Sci. 18*, 1015.
Dunn, R. (1981) Ph.D. Thesis. University of Essex, U.K.
Eagles, C. F. (1969) *Ann. Bot. 33*, 937.
Eagles, C. F. & Treharne, K. J. (1978) *Photosynthetica 3*, 29.
Earl, D. E. (1975) *Forest energy and economic development.* Clarendon Press.
Eckardt, F. E. (1975) In: *Photosynthesis and productivity in different environments.* (J. P. Cooper, ed.) p. 173. Cambridge University Press.
Eckardt, F. E., Heim, G., Methy, M., Saugier, B. & Sauvezon, R. (1971) *Oecol. Plant. 6*, 51.
Edwards, G. E. & Walker, D. A. (1983) C_3, C_4 *mechanisms, cellular and environmental regulation of photosynthesis.* 552 pp. Blackwell.
Ehleringer, J. (1978) *Oecologia 31*, 255.
Ehleringer, J. & Björkman, O. (1977) *Plant Physiol. 59*, 86.
Ehleringer, J. & Björkman, O. (1978) *Plant Physiol. 62*, 185.
Eik, K. & Hanway, J. J. (1966) *Agron. J. 58*, 16.
Eisbrenner, G. & Evans, H. J. (1983) *Ann. Rev. Plant Physiol. 34*, 105.
Ela, S. W., Anderson, M. A. & Brill, W. J. (1982) *Plant Physiol. 70*, 1564.
Ellis, R. P., Vogel, J. C. & Fuls, A. (1980) *S. Afr. J. Sci. 76*, 307.
El-Sharkawy, M. A., Loomis, R. S. Williams, W. A. (1968) *J. Appl. Ecol. 5*, 243.
Enoch, H. Z. (1977) *Agric. Met. 18*, 373.
Enoch, H. Z. & Kimbal, B. A. eds. (1984) *Carbon Dioxide Enrichment of Greenhouse Crops* C.R.C. Press, Boca Raton, Florida. (In press)
Epstein, E. (1972) *Mineral nutrition of plants: Principles and perspectives.* 412 pp. Wiley.
Epstein, E., & Norlyn, J. D. (1977) *Science 197*, 249.

REFERENCES

Evans, G. C. & Coombe, D. E. (1959) *J. Ecol. 47*, 103.
Evans, L. T., ed. (1975) Crop physiology. 374 pp. Cambridge University Press.
Evans, L. T. & Dunstone, R. L. (1970) *Aust. J. Biol. Sci. 23*, 725.
Evans, L. T. & Wardlaw, I. F. (1976) *Adv. Agron. 28*, 301.
F.A.O. Commodity Projections 1975-1985 (1979) F.A.O. Rome.
Farquhar, G. D. (1979) *Arch. Biochem. Biophys. 193*, 456.
Farquhar, G. D. & Caemmerer, von S. (1982) In: *Encycl. Plant Physiol. N. S. Vol. 12B* (O. L. Lange, P. S. Nobel, C. B. Osmond, H. Ziegler, eds.), p. 549. Springer-Verlag.
Farquhar, G. D., Caemmerer, von S., & Berry, J. A. (1980) Planta 149, 78.
Farquhar, G. D. & Sharkey, T. D. (1982) *Ann. Rev. Plant Physiol. 33*, 317.
Feller, W. & Erismann, K. H. (1978) *Z. Pflanzenphysiol. 90*, 235.
Fellows, R. J. & Boyer, J. S. (1978) *Protoplasma 93*, 381.
Fick, G. W., Loomis, R. S. & Williams, W. A. (1975) *Crop Physiology* (L. T. Evans, ed.) p. 259. Cambridge University Press.
Fick, G. W., Williams, W. A. & Loomis, R. S. (1973) *Crop Sci. 13*, 413.
Fischer, K., Kramer, D. & Ziegler, H. (1973) *Protoplasma. 76*, 83.
Fischer, R. A. (1973) In: *Plant response to climatic factors* (R. O. Slatyer, ed.) p. 233. Unesco, Paris.
Fischer, R. A. (1980) In: *Adaptation of plants to water and high temperature stress* (N. C. Turner & P. J. Kramer, eds.) p. 323. Wiley-Interscience.
Fischer, R. A., I. Aguilar, R. Maurero, & S. Rivasa (1976) *J. Agric. Sci. 87*, 137.
Fischer, R. A., Hsaio, T. C. & Hagan, R. M. (1970) *J. Exp. Bot. 21*, 371.
Fischer, R. A. & Turner, N. C. (1978) *Ann. Rev. Plant Physiol. 29*, 277.
Flowers, T. J., Troke, P. F. & Yeo, A. R. (1977). *Ann. Rev. Plant Physiol. 28*, 89.
Fowler, D. & Cape, J. N. (1982) In: *Effects of gaseous air pollution in agriculture and horticulture* (M. H. Unsworth & D. P. Ormrod, eds.) p. 3. Butterworth.
Fox, F. M. & Caldwell, M. M. (1978) *Oecologia 36*, 173.
Foyer, C. H. & Hall, D. O. (1980) *Trends Biochem. Sci. 5*, 188.
Freyer, H. D. (1979) *The global carbon cycle* (B. Bolin, E. T. Degens, S. Kempe, P. Ketner, eds.) p. 79, Wiley.
Fry, K. E. (1970) *Plant Physiol. 75*, 465.
Fuchs, M. (1972) *Optimizing the Soil Physical Environment Towards Greater Crop Yields* (D. Hillel, ed.) p. 173. Academic Press.
Fuchs, M., Schulze, E. D. & Fuchs, M. I. (1977) *Oecologia 29*, 329.
Fuchs, M. & Stanhill, G. (1980) *Plant Cell Env. 3*, 175.
Fukai, S. & Loomis, R. W. (1976) *Agric. Meteorol. 17*, 353.
Gaastra, P. (1959) *Meded., Landbhoogesch. Wageningen 59*, 1.
Gaastra, P. (1965) *Environmental Control of plant growth* (L. T. Evans, ed.) p. 113. Academic Press.
Gale, J. (1975) In: *Ecol. Stud. Vol. 15* (A. Poljakoff-Mayber, J. Gale, eds.) p. 168. Springer-Verlag.
Gallagher, J. N. & Biscoe, P. V. (1978) *J. Agric. Sci. 91*, 47.
Gallagher, J. N., Biscoe, P. V. & Scott, R. K. (1975) *J. Appl. Ecol. 12*, 319.
Gates, D. M. (1965) *Met. Monograph. 6*, 1.
Gausmann, H. W., Allen, W. A., Wiegand, C. L. Escobas, D. E., Rodrigues, R. E., & Richardson, A. J. (1973) *U.S.D.A. Tech. Bull.* 1465.
Gerakis, P. A. & Papakosta-Taspoula, D. (1980) *Agric. Met. 21*, 129.
Gerwick, B. C. (1982) In: *CRC Handbook of biosolar resources Vol. 1 Part 2* (A. Mitsui, C. C. Black, eds.) p. 213. CRC Press, Boca Raton.
Gezelius, K. & Hallén, M. (1980) *Physiol. Plant. 48*, 88.
Gezelius, K. & Hällgren, J.-E. (1980) *Physiol. Plant. 49*, 354.
Gibbs, M. & Latzko, E. (eds.) (1979) *Encycl. Plant Physiol. N. S. Vol. 6* 578 pp. Springer-Verlag.
Gifford, R. M. (1977) *Aus. J. Plant Physiol. 4*, 99.

Gifford, R. M. (1979) *Aust. J. Plant Physiol. 6,* 367.
Gifford, R. M., & Evans, L. T. (1981) *Ann. Rev. Plant Physiol. 32,* 485.
Giles, K. R., Beardsell, M. F. & Cohen, D. (1974) *Plant Physiol. 54,* 208.
Glinka, Z. & Katchansky, M. Y. (1970) *Isr. J. Bot. 19,* 533.
Godzik, S. & Sassen, M. M. A. (1974) *Phytopathol. Z. 79,* 155.
Golbeck, J. H., Lien, S. & San Pietro, A. (1977) In: *Encycl. Plant Physiol. N.S. Vol. 5* (A. Trebst and M. Avron, eds.) p. 94. Springer-Verlag.
Good, N. E. & Bell, D. H. (1980) *The biology of crop productivity* (P. S. Carlson, ed.) p. 3. Academic Press.
Goudriaan, J. (1977) *Crop Micrometeorology: a Simulation Study.* 250 pp. Wageningen.
Goudriaan, J. & Atjay, Jr. G. L. (1979) *The global carbon cycle* (B. Bolin, E. T. Degens, S. Kempe, & P. Ketner, eds.) p. 237. Wiley.
Govindjee ed. (1983) *Photosynthesis Vol. 1* 799 pp. Academic Press.
Grace, J. & Russell, G. (1977) *J. Exp. Bot. 28,* 268.
Grahl, H. & Wild, A. (1975) In: *Environmental and biological control of photosynthesis* (R. Marcelle, ed.) p. 107. Junk.
Greenway, H. (1973) *J. Aust. Inst. Agric. Sci. 39,* 24.
Greenway, H. & Munns, R. (1980) *Ann. Rev. Plant Physiol. 31,* 141.
Greenway, H. & Osmond, C. B. (1972) *Plant Physiol. 49,* 256.
Grennfelt, P. (1981) *Report Eur. Conf. Acid Rain* p. 31. Goteborg.
Grime, J. P. & Hunt, R. (1975) *J. Ecol. 63,* 393.
Gross, K. (1976) *Forstwiss. Centrabl. 95,* ,211.
Guerrero, M. G., Vega, J. M. & Losada, M. (1981) *Ann. Rev. Plant Physiol. 32,* 169.
Haberlandt, G. (1884) *Physiological Plant Anatomy* (Transl. M. Drummond) Macmillan.
Hall, A. E. (1971) *Carn. Inst. Wash. Yearbk. 70,* 530.
Haenhel, W. (1984) *Ann. Rev. Plant Physiol. 35,* 659.
Hall, A. E. & Björkman, O. (1975) *Perspectives of Biophysical Ecology* (D. M. Gates & R. B. Schmerl) p. 55. Springer.
Hall, D. O. (1976) In: *The Intact Chloroplast* (Barber, J. Ed.) p. 135. Elsevier.
Hall, D. O. (1979) *Solar Energy 22,* 307.
Hall, D. O. (1983) In: *Energy from biomass* (A. Strube, P. Chartier & G. Schleser, eds.) p. 43. Applied Science Publ.
Hall, D. O., Barnard, G. W. & Moss, P. A. (1982) *Biomass for energy in the developing countries.* Pergamon Press.
Hällgren, J. E. (1978) *Sulfur in the environment Part II* (J. O. Nriagu, ed.) p. 164. Wiley.
Hällgren, J. E. (1980) *Swedish coniferous project. Tech. Rep. 25* (S. Linder, ed.) p. 125. Swedish University of Agricultural Sciences, Uppsala.
Halliwell, B. (1981) *Chloroplast metabolism* 257 pp. Oxford University Press.
Halliwell, B. (1982a) *Superoxide dismutase* (L. W. Oberley, ed.) p. 89. CRC Press.
Halliwell, B. (1982b) *Trends Biochem. Sci. 1982.,* 290.
Hampp, R., Beulish, K. & Ziegler, M. (1976) *Z. Pflanzenphysiol. 77,* 336.
Hanks, R. J., Gardner, H. R. & Florian, R. L. .(1969) *Agron. J. 61,* 30.
Hanson, W. D. (1971) *Crop Sci. 11,* 334.
Hardy, R. W. F. & Havelka, U. D. (1976) *Symbiotic nitrogen fixation in plants.* (P. Nutman, ed.). p. 421. Cambridge University Press.
Hardy, R. W. F. & Havelka, U. D. (1977) In: *Biological solar energy conversion* (A. Mitsui, S. Miyashi, A. San Pietro, S. Tamura, eds.) p. 299. Academic Press.
Hardy R. W. F., Havelka, U. D. & Quebedeaux, B. (1978) In: *Proc. 4th Int. Phot. Cong.* (D. O. Hall, J. Coombs, T. W. Goodwin, eds.) p;. 695. The Biochemical Society, London.
Harper, J. L. (1963) *Nature 197,* 917.
Hartsock, T. L. & Nobel, P. S. (1976) *Nature 262,* 574.
Hasson, E., Poljakoff-Mayber, A. & Gale, J. (1983) In: *Effects of stress on photosynthesis* (R. Marcelle, H. Clijsters, J. Gale, eds.) p. 305. Junk.

REFERENCES

Hatch, M. D. (1977) *Plant Cell Physiol. 3*, 311.
Hatch, M. D. (1982) In: *C.R.C. Handbook of biosolar resources. Vol. 1. Part 1.* (A Mitsui, C. C. Black, eds.) p. 185. CRC Press, Boca Raton.
Hatch, M. D. & Boardman, N. K. (1981) *The biochemistry of plants. Vol. 8* 521 pp. Academic Press.
Hatch, M. D. & Kagawa, T. (1970) *Ann. Rev. Plant Physiol. 21*, 142.
Hatch, M. D. & Kagawa, T. (1973) *Arch. Biochem. Biophys. 159*, 842.
Hatch, M. D., Kagawa, T. & Craig, S. (1975) *Aust. J. Plant Physiol. 2*, 111.
Hatch, M. D. & Osmond, C. B. (1976) In: *Encycl. Plant Physiology N. S. Vol. 3* (C. R. Stocking, W. Heber, eds.) p. 144. Springer-Verlag.
Hatch, M. D. Slack, C. R. (1966) *Biochem. J. 101*, 103.
Hatch, M. D. & Slack, C. R. (1970) *Ann. Rev. Plant Physiol. 21*, 141.
Hatch, M. D., Slack, C. R. & Bull, T. A. (1969) *Phytochemistry 8*, 697.
Hattersley, P. W. & Watson, L. (1975) *Phytomorphology 25*, 325.
Hattersley, P. W. & Watson, L. (1976) *Aust. J. Bot. 24*, 297.
Haworth, P., Kylle, D., Horton, P. & Arntzen, C. J. (1982) *Photochem. Photobiol. 36*, 743.
Hayashi, K. (1966) *Proc. Crop Sci. Soc. Japan 34*, 205.
Haystead, A. & Sprent, J. I. (1981) In: *Physiological processes limiting plant productivity* (C. B. Johnson, ed.) p. 345. Butterworths.
Heath, R. L. (1980) *Ann. Rev. Plant Physiol. 31*, 395.
Heichel, G. H. & Musgrave, R. B. (1969) *Crop Sci. 9*, 483.
Heldt, H. W. (1979) In: *Encycl. Plant Physiol. N. S. Vol. 6.* (Gibbs, M., Latzko, E. eds.) p. 202. Springer-Verlag.
Heldt, H. W. (1981) In: *Proc. 5th Int. Phot. Cong. Vol. 4.* (G. Akoyunoglou, ed.) p. 213. Balaban International, Philadelphia.
Hesketh, J. D. & Baker, D. N. (1967) *Crop Sci. 7*, 285.
Hesketh, J. D. & Jones, R. W. (1980) *Predicting photosynthesis for ecosystem models. Vol. 1*, 273 pp. C.R.C. Press, Boca Raton.
Hickelton, P. R. & Oechel, W. C. (1976) *Can. J. Bot. 54*, 1104.
Highkin, H. R., Boardman, N. K. & Goodchild, D. J. (1969) *Plant Physiol. 44*, 1310.
Hill, A. C. & Littlefield, N. (1969) *Environ. Sci. Technol. 3*, 52.
Holdgate, M. W., Kassas, M. & White, G. F. (eds.) (1982) *The world environment UNEP Report.* 677 pp. Tycooly International.
Holland, H. D. (1978) *The Chemistry of the Atmosphere and Oceans*, 351 pp. Wiley.
Hoveland, C. S., Foutch, H. W. & Buchanan, G. A. (1974) *Agron. J. 66*, 686.
Hsiao, T. C. (1973) *Ann. Rev. Plant. Physiol. 24*, 519.
Huang, C.-Y., Bazzaz, F. A. & Vanderhoef, L. N. (1974) *Plant Physiol. 54*, 122.
Huffaker, R. C., Radin, T., Kleinkopf, G. E. & Cox, E. L. (1970) *Crop Sci. 10*, 471.
Hughes, A. P. & Freeman, P. R. (1967) *J. Appl. Ecol. 4*, 553.
Huner, N. P. R. & MacDowell, F. D. H. (1979) *Can. J. Biochem. 57*, 1036.
Hunt, R. (1978) *Plant Growth Analysis Studies in Biology No. 96*, 67 pp. Arnold.
Hunt, R. & Parsons, I. T. (1974) *J. Appl. Ecol. 11*, 297.
Hunt, R. & Parsons, I. T. (1977) *J. Appl. Ecol. 14*, 965.
Hunt, W. F. & Loomis, R. S. (1979) *Ann. Bot. 44*, 5.
Hurd, R. G.(1977) *Ann. Bot. 41*, 779.
Hussey, A. & Long, S. P. (1982) *J. Ecol. 70*, 757.
Imbamba, S. K. & Papa, G. (1979) *Photosynthetica 13*, 315.
Incoll, L. D. (1977) In: *Environmental effects on Crop Physiology* (J. J. Landsberg & C. V. Cutting, eds.) p. 137. Academic Press.
Incoll, L. D. & Wright, W. H. (1969) *Conn. Agr. Exp. Sta. Bull. Soils, 15*, 1.
Inoue, E., Uchijima, Z., Udagawa, T., Horie, T. & Kobayashi, K. (1968) *J. Agric. Met. 23*, 165.
Isabe, S. (1962) *Bull. Nat. Inst. Agric. Sci. Tokyo. A9.* 29.
Ishii, R., Sarnejima, M. & Murata, Y. (1977) *Jap. J. Crop Sci. 46*, 97.

Ivory, D. A. & Whiteman, P. C. (1978) *Aust. J. Plant. Physiol. 5,* 149.
Iwata, F. & Okubo, T. (1971) *Proc. Crop Sci. Soc. Japan 40,* 262.
Jacques, W. A. & Schwass, R. L. (1956) *N.Z. J. Sci. Tech. A 37,* 569.
Jacobson, J. W. & Hill, A. C. (1970) *Recognition of air pollution injury to vegetation: a pictorial atlas.* Air pollution control association, Pittsburgh.
Jagendorf, A. T. (1977) In: *Encycl. Plant Physiol. N. S. Vol. 5* (Trebst, A., Avron, M. eds.) p. 307. Springer-Verlag.
Jarvis, P. G. (1970) *Prediction and measurement of photosynthetic productivity* (I. Setlik, ed.) p. 353. Wageningen.
Jarvis, P. G. (1971) In: *Plant Photosynthetic Production.* (Sestak, Z., Catsky, J., Jarvis, P. G. eds.) p. 566. Junk.
Jarvis, P. G., James, G. B. & Landsberg, J. J. (1976) In: *Vegetation and the atmosphere. Vol. 2* (J. L. Monteith, ed.) p. 171. Academic Press.
Jarvis, P. G. & Jarvis, M. S. (1964) *Physiol. Plant. 17,* 65.
Jeffree, C. (1976) *Costs of damage to vegetation from SO_2 pollution.* Unpublished report to Dept. of Environment, U.K. Dept. of Forestry & Natural Resources, University of Edinburgh.
Jennings, D. H. (1968) *New Phytol, 67,* 899.
Jensen, K. F. (1975) *USDA Forest Service Res. Note. NE-209.*
Jensen, N. F. (1978) *Science 201,* 317.
Johnson, R. R., Frey, N. N. & Moss, D. N. (1974) *Crop Sci. 14,* 728.
Jones, H. G. (1973) *New Phytol. 72,* 1095.
Jones, H. G. & Fanjul, L. (1983) In: *Effects of stress on photosynthesis* (R. Marcelle, H. Clijster, M. van Poucke, eds.) p. 75. Junk.
Jones, R. J., Nelson, C. J. & Sleper, D. A. (1979) *Crop Sci. 19,* 631.
Jørgensen, S. E. (1979). *Handbook of environmental data and ecological parameters.* Pergamon Press.
Joshi, G. V. (1976) *Studies in Photosynthesis under Saline conditions.* 195 pp. Shivaji University, Kolhapur.
Junge, W. (1977) In: *Encycl. Plant Physiol. N. S. Vol. 5* (A. Trebst, M. Avron, eds.) p. 59. Springer-Verlag.
Kanai, R. & Kashiwaga, M. (1975) *Plant Cell Physiol. 16,* 669.
Kasanaga, H. & Monsi, M. (1954) *Jap. J. Bot. 14,* 304.
Katie, T. (1970) *Soil Sci. Plant Nutr. 15,* 245.
Kawashima, R. (1980) *Green Energy Program.* Agriculture, Forestry & Fisheries Research Council Secretariat, Tokyo, Japan.
Keck, R. W. & Boyer, J. S. (1974) *Plant Physiol. 53,* 474.
Kelly, G. J., Latzko, E. & Gibbs, M. (1975) *Ann. Rev. Plant Physiol. 27,* 181.
Kennedy, R. A. & Laetsch, W. M. (1974) *Science 184,* 1087.
Kerr, R. A. (1980) *Science 208,* 1358.
Kestler, D. P., Mayne, B. C., Ray, T. B., Goldstein, L. D., Brown, R. H. & Black, C. C. (1975) *Biochem. Biophys. Res. Comm. 66,* 1439.
Keys, A. J., Sampaio, E. V. S. B., Cornelius, M. J. & Bird, I. F. (1977) *J. Exp. Bot. 28,* 525.
Keys, A. J. & Whittingham, C. P. (1981) In: *Physiological processes limiting plant productivity* (C. B. Johnson, ed.) p. 137. Butterworths.
King, R. W. & Evans, L. T. (1967) *Aust. J. Biol. Sci. 20,* 623.
Kitajima, M. & Butler, W. L. (1975) *Biochem. Biophys. Acta. 408,* 297.
Klein, R. M. (1978) *Biol. Rev. 44,* 1.
Kleshin, A. F. & Shuglin, I. A. (1959) *Dokl. Acad. SSR 125,* 1158. (in Russian).
Kluge, M. (1979) In: *Encycl. Plant Physiol. N. S. Vol. 6* (M. Gibbs, E. Latzko, eds.) p. 113. Springer-Verlag.
Kluge, M., Fischer, A. & Buchanan-Bollig, I. C. (1982) In: *Crassulacean Acid Metabolism* (I. P. Ting & M. Gibbs, eds.) p. 31. American Society of Plant Physiologists, Rockville.
Knox, R. S. (1977) In: *The intact chloroplast. Vol. 2* (J. Barber, ed.) p. 55. Elsevier.

REFERENCES

Koch, W., Klein, E. & Walz, H. (1968) *Siemens, Z. 42*, 392.
Koiwai, A. & Kisaki, T. (1976) *Plant Cell Physiol 17:* 1199.
Körner, C., Scheel, J. A. & Bauer, H. (1979) *Photosynthetrica 13*, 45.
Kortschak, H. P., Hartt, C. E. & Burt, C. D. (1965) *Plant Physiol. 40*, 209.
Koziol, M. J. & Jordan, C. F. (1979) *J. Exp. Bot. 29*, 1037.
Kramer, P. (1980) *Linking Research to Crop Production* (R. C. Staples, R. J. Kuhr, eds.) p. 51. Plenum Press.
Krause, G. M. & Santarius, K. A. (1975) *Planta 127*, 285.
Kriedemann, P. E. & Downton, W. J. S. (1981) In: *The physiology and biochemistry of drought resistance in plants* (L. G. Paleg & D. Aspinall, ed.) p. 283. Academic Press.
Kriedemann, P. E. & Loveys, B. R. (1974) *R. Soc. N.Z. Bull. 12*, 461.
Krinsky, N. I. (1971) *Carotenoids* (O. Ister, ed.) p. 669. Birkhäuser, Basel.
Ku, S. B. & Edwards, G. E. (1978) *Planta 140*, 1.
Ku, S. B. & Hunt, L. A. (1977) *Can. J. Bot. 55*, 872.
Kuroiwa, S. (1969) In: *Prediction and measurement of photosynthetic productivity* (I. Setlik, ed.) p. 79. Wageningen.
Kuroiwa, S. & Monsi, M. (1963) *J. Agric. Met. 18*, 143.
Laetsch, W. M. (1969) *Sci. Prog. 57*; 323.
Laetsch, W. M. (1974) *Ann. Rev. Plant Physiol. 25*, 27.
Laing, W. A., Ogren, W. L. & Hageman, R. H. (1974) *Plant Physiol. 54*, 678.
Lamborg, M. R. (1980) *Linking research to crop production* (R. C. Staples, R. J. Kuhr, eds.) p. 115. Plenum Press.
Lamoreaux, R. J. & Chaney, W. R. (1978) *Env. Pollut. 17*, 259.
Landsberg, J. J. et al. (1975) *J. Appl. Ecol. 12*, 659.
Lange, O. L., Schulze, E-D., Evenari, M., Kappen, L. & Buschbom, U. (1974) *Oecologia 17*, 97.
Lange, O. L., Schulze, E-D., Evenari, M., Kappen, L. & Buschbom, U. (1975) *Oecologia 18*, 45.
Lange, O. L., Schulze, E-D., Evenari, M., Kappen, L. & Buschbom, U. (1978) *Oecologia 34*, 89
Lapina, L. P. & Bikhukhometova, S. A. (1972) *Soviet Plant Physiol. 19*, 792.
Larcher, W. (1980) *Physiological Plant Ecology* 303 pp. (Springer-Verlag, Publ.).
Larcher, W. (1981) In: *Physiological processes limiting plant productivity* (Johnson, C. B., ed.) p. 253. Butterworths.
LaRue, T. A. G. & Kurz, W. G. W. (1973) *Can. J. Microbiol. 19*, 304.
Latimore, M., Gidens, J. & Ashley, D. A. (1977) *Crop Sci. 17*, 399.
Latzko, E. & Kelly, G. J. (1979) In: *Encycl. Plant Physiol. N. S. Vol. 6* (Gibbs, M., Latzko, E. eds.) p. 239. Springer-Verlag.
Law, R. M. & Mansfield, T. A. (1982) In: *Effects of gaseous air pollution in agriculture and horticulture* (M. H. Unsworth & D. P. Ormrod, eds.) p. 93. Butterworths.
Lawlor, D. W. (1976a) *Photosynthetica 10*, 378.
Lawlor, D. W. (1976b) *Photosynthetica 10*, 431.
Lawlor, D. W. (1979) *Stress physiology in crop plants.* (H. Mussel, R. C. Staples, eds.) p. 303. Wiley.
Lawlor, D. W. & Fock, H. (1977) *J. Exp. Bot. 28*, 329.
Lawlor, D. W., Mahon, J. D. & Fock, H. (1977) *Can. J. Bot. 11*, 322.
Lea, P. J. & Miflin, B. J. (1979) In: *Encycl. Plant Physiol. N. S. Vol. 6* (M. Gibbs, E. Latzko, eds.) p. 445. Springer-Verlag.
Ledent, J. F. (1978) *Ann. Bot. 42*, 345.
Lee, K. C., Campbell, R. W. & Paulsen, G. M. (1974) *Crop Sci. 14*, 279.
Leegood, R. C. & Walker, D. A. (1982) *Planta 156*, 449.
Legg, B. J., Day, W., Lawlor, D. W. & Parkinson, K. J. (1979) *J. Agric. Sci. 92*, 703.
Lemon, E. (1963) In: *Environmental control of plant growth* (L. T. Evans, ed.) p. 55. Academic Press.

Lemon, E. (1965) In: *Plant Environment and Efficient Water Use* p. 28. ASA monograph, Ames, Iowa.
Lemon, E. (1967) *Harvesting the sun* (A. San Pietro, F. A. Greer & T. J. Army, eds.) p. 263. Academic Press.
Lemon, E. ed. (1984) CO_2 *and Plants: The Response of Plants to Rising Levels of Atmospheric Carbon Dioxide.* Westview Press, Colorado. (In press.)
Leverenz, J. W. & Jarvis, P. G. (1979) *J. Appl. Ecol. 16,* 919.
Levin, D. A. (1976) *Am. Nat. 110,* 261.
Lieth, H. & Whittaker, R. H. (1975) *The primary production of the biosphere* 339 pp. Springer-Verlag.
Linden, C. D., Wright, K. L., McConnell, H. M. & Fox, C. F. (1973) *Proc. Natl. Acad. Sci. 70,* 2271.
Linthurst, R. A. & Reimold, R. J. (1978) *J. Appl. Ecol. 15,* 919.
Littleton, E. J. (1971) Ph.D. Thesis, Univ. of Nottingham.
Lloyd, N. D. H. & Canvin, D. T. (1977) *Can. J. Bot. 55,* 3006.
Lloyd, N. D. H. & Woolhouse, H. W. (1976) *New Phytol.* 77, 553.
Lommen, P. W., Schwintzer, C. R., Yocum, C. S. & Gates, D. M. (1971) *Planta 98,* 195.
Long, S. P. (1982) In: *Techniques in bioproductivity and photosynthesis* (J. Coombs, D. O. Hall, eds.) p. 25. Pergamon Press.
Long, S. P. (1983) *Plant Cell Environ., 6,* 345.
Long, S. P., East, T. M. & Baker, N. R. (1983) *J. Exp. Bot. 34,* 177.
Long, S. P. & Incoll, L. D. (1979) *J. Appl. Ecol. 16,* 879.
Long, S. P., Incoll, L. D. & Woolhouse, H. W. (1975) *Nature 257,* 622.
Long, S. P. & Woolhouse, H. W. (1978) *J. Exp. Bot. 29,* 803.
Longstreth, D. J. & Nobel, P. S. (1979) *Plant Physiol. 63,* 700.
Longstreth, D. J. & Nobel, P. S. (1980) *Plant Physiol. 65,* 541.
Loomis, R. S. (1963) *Crop Sci. 3,* 67.
Loomis, R. S. (1976) *Sci. Am. 235 (3),* 99.
Loomis, R. S. & Gerakis, P. A. (1975) In: *Photosynthesis and productivity in different environments* (J. P. Cooper, ed.) p. 145. Cambridge University Press.
Loomis, R. S., Rabbinge, R. & Ng, E. (1979) *Ann. Rev. Plant Physiol. 30,* 339.
Loomis, R. S. & Williams, W. A. (1963) *Crop Sci. 3,* 67.
Loomis, R. S. & Williams, W. A. (1969) *Physiological aspects of Crop Yield* (J. Eastin, F. A. Haskins, C. Y. Sullivan & C. H. M. Van Bavel, eds.) p. 27. American Society of Agronomy and Crop Science, Madison, Wisconsin.
Loomis, R. S., Williams, W. A. & Hall, A. E. (1971) *Ann. Rev. Plant Physiol. 22,* 431.
Lorimer, G. H. (1981) *Ann. Rev. Plant Physiol. 32,* 349.
Lougham, B. C. (1964) *Agrochimica 8,* 189.
Loveys, B. R. & Kriedemann, P. E. (1973) *Physiol. Plant. 28,* 476.
Ludlow, M. M. (1975) *Environmental and Biological Control of Photosynthesis* (R. Marcelle, ed.) p. 123. Junk.
Ludlow, M.M. (1976) *Water and Plant Life. Ecological Studies. Vol. 19* (Lange, O. L., Kappen, L. and Schulze, E. D. eds.) p. 364. Springer.
Ludlow, M. M. (1980) *Photosyn. Res. 1,* 243.
Ludlow, M. M. (1982) In: *Techniques in photosynthesis and bioproductivity* (J. Coombs & D. O. Hall eds.) p. 5. Pergamon Press.
Ludlow, M. M. & Ng, T. T. (1974) *Plant Sci. Lett. 33,* 235.
Ludlow, M. M. & Ng, T. T. (1976) *Aust. J. Plant Physiol. 3,* 401.
Ludlow, M. M. & Wilson, G. L. (1971a) *Aust. J. Biol. Sci. 24,* 449.
Ludlow, M. M. & Wilson, G. L. (1971b) *Aust. J. Biol. Sci. 24,* 1065.
Ludlow, M. M. & Wilson, G. L. (1971c) *Aust. J. Biol. Sci. 24,* 1077.
Ludwig, L. J. & Canvin, D. T. (1971) *Can. J. Bot. 49,* 1299.
Ludwig, L. J., Charles-Edwards, D. A. & Withers, A. C. (1975) *Environmental and biological control of photosynthesis* (R. Marcelle, ed.) p. 29. Junk.

REFERENCES

Ludwig, L. J., Saeki, T. & Evans, L. T. (1965) *Aust. J. Biol. Sci. 18*, 1103.
Lyons, J. M. & Breidenbach, R. W. (1979) In: *Stress physiology in crop plants*. (M. Mussel & R. C. Staples eds.) p. 179. Wiley.
Maas, E. U. & Hoffman, G. J. (1977) *J. Irr. Drain Div. 103*, 115.
Mahon, J. D., Fock, M., Hohler, T. & Canvin, D. T. (1974) *Planta 120*, 113.
Malhotra, S. S. (1977) *New Phytol. 78*, 101.
Malkin, R. (1982) *Ann. Rev. Plant Physiol. 33*, 455.
Mansfield, T. A. & Wilson, J. A. (1981) In: *Physiological processes limiting plant productivity* (C. B. Johnson, ed.) p. 237. Butterworths.
Mantai, K. E., Wong, J. & Bishop, N. I. (1970) *Biochim. Biophys. Acta 197*, 257.
Martin, B., Martensson, O. & Öquist, G. (1978) *Physiol. Plant. 44*, 102.
Martin, B., Ort, D. R. & Boyer, J. S. (1981) *Plant Physiol. 68*, 729.
Mason, C. F. & Bryant, R. J. (1973) *J. Ecol. 63*, 71.
Maudlin, W. P. (1980) *Science 209*, 148.
McArthur, J. A., Hesketh, J. D. & Baker, D. N. (1975) *Crop physiology* (L. T. Evans, ed.) p. 297. Cambridge University Press.
McCarty, R. E. & Portis, A. R. (1976) *Biochemistry 15*, 5110.
McCree, K. J. (1972) *Agric. Met. 9*, 191.
McCree, K. J. (1981) In: *Encycl. Plant Physiol. N. S. Vol. 12A* (O. L. Lange, P. S. Nobel, C. B. Osmond, H. Ziegler, eds.) p. 41. Springer-Verlag.
McCree, K. J. & Troughton, J. H. (1966) *Plant Physiol. 41*, 1615.
McPherson, H. G. & Boyer, J. S. (1977) *Agron. J. 69*, 714.
McPherson, H. G. & Slatyer, R. O. (1973) *Aust. J. Biol. Sci. 26*, 329.
McWilliam, J. R. & Ferrar, P. J. (1974) *Mechanisms of regulation of plant growth* (R. L. Bieleski, A. R. Ferguson, M. M. Cresswell, eds.) p. 467. Royal Society of New Zealand, Wellington.
McWilliam, J. R. & Naylor, A. W. (1967) *Plant Physiol. 42*, 1711.
Mederski, M. J., Chen, L. M. & Curry, R. B. (1975) *Plant Physiol. 55*, 589.
Medina, E., Delgado, M., Troughton, J. H. & Medina, J. D. (1977) *Flora 166*, 137.
Metevier, J. R. & Dale, J. E. (1977) *Ann. Bot. 41*, 1287.
Miflin, B. J. (1980) In: *The biology of crop productivity* (P. S. Carlson, ed.) p. 255. Academic Press.
Milner, C. & Hughes, R. E. (1968) *Methods for the measurement of primary production of grassland*. 70 pp. Blackwell.
Minchin, F. R. & Pate, J. S. (1973) *J. Exp. Bot. 24*, 259.
Mohammed, S. (1978) *Workshop on membrane biophysics and development of salt tolerant plants*. University of Agriculture, Faisalabad, Pakistan.
Mohanty, P. & Boyer, J. S. (1976) *Plant Physiol. 57*, 704.
Molina, J. M. & Rowland, F. S. (1974) *Nature, 249*, 810.
Monsi, M. & Saeki, T. (1953) *Jap. J. Bot. 14*, 22.
Monson, R. K., Littlejohn, R. O. & Williams, G. J. (1982) *Photosyn. Res. 3*, 153.
Monson, R. K., Stidman, M. A., Williams, G. J., Edwards, G. E. & Uribe, E. G. (1982). *Plant Physiol. 69*, 921.
Monteith, J. L. (1962) *Neth. J. Agric. Sci. 10*, 334.
Monteith, J. L. (1965a) *Ann. Bot. 29*, 17.
Monteith, J. L. (1965b) *Field Crop Abstr. 18*, 213.
Monteith, J. L. (1969) *Physiological Aspects of Crop Yield* (J. D. Eastin, F. A. Haskins, C. Y. Sullivan, C. H. Van Bavel, eds.) p. 89. American Society of Agronomy and Crop Science, Madison, Wisconsin.
Monteith, J. L. (1973) *Principles of environmental physics*. 241 pp. Arnold.
Montieth, J. L. (1976) *Vegetation and the Atmosphere Vols 1 & 2*. 706 pp. Academic Press.
Monteith, J. L. (1977) *Phil. Trans. Roy. Soc. B. 281*, 277.
Monteith, J. L. (1978) *Expl. Agric. 14*, 1.
Monteith, J. L. (1981) In: *Physiological processes limiting plant productivity* (C. B. Johnson,

ed.) p. 23. Butterworths.
Monteith, J. L. & Sziecz, G. (1960) *Quart. J. Roy. Met. Soc. 86*, 205.
Montfort, C. (1950) *Zeit. Naturforsch. 56*, 221.
Mooney, H. A. (1978) *Oecologia, 36*, 103.
Mooney, H. A. & Harrison, A. T. (1970) In: *Prediction and measurement of photosynthetic productivity* (C. T. de Wit, ed.) p. 411. Wageningen.
Mooney, H. A., Björkman, O. & Collatz, G. J. (1977) *Carn. Inst. Wash. Yearbk 76*, 328.
Moore, A. L. (1977) *Nature, 267*, 307.
Moore, B., Boone, R. D., Hobbie, J. E., Houghton, R. A., Melillo, J. M., Peterson, B. J., Shaver, G. R., Vorosmarty, C. J. & Woodwell, G. M. (1981) *In Carbon Cycle Modelling* (B. Bolin, ed.) p. 365. Wiley.
Moore, P. D. (1981) *New Scientist 89*, 394.
Moore, P. D. (1983) *J. Geol. Sci. 140*, 13.
Morgan, J. A. & Brown, R. H. (1979) *Plant Physiol. 64*, 257.
Morgan, J. A., Brown, R. H. & Reger, B. J. (1980) *Plant Physiol. 65*, 156.
Morrison, S. L., Huffaker, R. C. & Loomis, R. S. (1979) *Plant Physiol. 63*, Suppl. 46.
Mühlethaler, K. (1977) In: *Encycl. Plant Physiol. N. S. Vol. 5* (A. Trebst, M. Avron, eds.) p. 503. Springer-Verlag.
Müller, R. N., Miller, J. E. & Sprugel, D. G. (1979) *J. Appl. Ecol. 16*, 567.
Murata, Y. (1975) *Agric. Met. 15*, 117.
Murata, Y. & Iyama J. (1963) *Proc. Crop Sci. Soc. Japan 32*, 315.
Murdie, P. J. (1974) In: *Ecology of halophytes* (R. J. Reimold, W. H. Queen, eds.) p. 565. Academic Press.
Murphy, C. E. & Knoerr, K. R. (1970) *Proc. 1970 Summer Computer Simulation Conf.* p. 786. Simultations Council Inc. La Jolla, Cal.
Murphy, C. E. & Knoerr, K. R. (1972) Eastern Deciduous Forest Biome Memo Report No. 72-10, p. 734, Oakridge National Laboratory, Tennessee.
Nakamura, H. & Saka, H. (1978) *Jap. J. Crop Sci. 47*, 707.
Nasyrov, Y. S. (1960) *Khlopchatnik (Cotton) 4*, 227. (in Russian.)
Nasyrov, Y. S. (1978) *Ann. Rev. Plant Physiol. 29*, 215.
National Academy of Sciences (1979) *Protection against depletion of stratospheric ozone by chlorofluorocarbons* Washington D.C.
Natr, L. (1970) *Physiol. Veg. 8*, 573.
Neales, T. F. & Incoll, L. D. (1968) *Bot. Rev. 34*, 107.
Neilson, R. E. (1977) *Photosynthetica 11*, 334.
Neilson, R. E., Ludlow, M. M. & Jarvis, P. G. (1972) *J. Appl. Ecol. 9*, 721.
Nelson, W. L. & Dale, R. F. (1978) *Agron. J. 70*, 734.
Newbould, P. J. (1967) *Methods for estimating the primary production of forests. IBP Handbook No. 2.* 62 pp. Blackwell.
Newton, J. E. & Blackman, G. E. (1970) *Ann. Bot. 34*, 329.
Ng, P. A. P. & Jarvis, P. G. (1980) *Plant Cell Env. 3*, 207.
Nicholls, A. O. & Calder, D. M. (1973) *New Phytol. 72*, 571.
Niebor, E., Richardson, D. H. S., Puckett, K. J. & Tomasini, F. D. (1976) *Effects of air pollutants on plants* (T. A. Mansfield, ed.) p. 61. Cambridge University Press.
Nir, I. & Poljakoff-Mayber, A. (1967) *Nature 213*, 418.
Nobel, P. S. (1974) *Biophysical Plant Physiology*, p. 488. Freeman.
Nobel, P. S. (1977) *Physiol. Plant.* 40, 137.
Nobel, P. S. (1980a) *Ecology 61*, 252.
Nobel, P. S. (1980b) In: *Adaptation of plants to water and high temperature stress* (N. C. Turner, P. J. Kramer, eds.) p. 43. Wiley-Interscience.
Nobel, P. S., Longstreth, D. J. & Hartsock, T. L. (1978) *Physiol. Plant 44*, 97.
Nobel, P. S. & Wang, C. T. (1973) *Arch. Biochem. Biophys. 157*, 338.
Noble, R. D. & Jensen, K. F. (1979) *Plant Physiol. 63 Suppl.*, 151.
Nobs, M. A. (1976) *Carn. Inst. Wash. Yearbk. 75*, 421.

REFERENCES

Nolan, W. G. & Smillie, R. M. (1976) *Biochim. Biophys, Acta 440*, 461.
Northcote, K. H. & Shene, J. K. M. (1972) *Australian Soils with Saline and Sodic Properties. Soil Publ. 27.* Commonwealth Science and Industrial Research Organisation, Melbourne.
Ogawa, T. (12975) *Physiol. Plant. 35*, 91.
Ogren, W. L. (1978) In: *Proc. 4th Int. Phot. Cong.* (D. O. Hall, J. Coombs, T. W. Goodwin, eds.) p. 721. The Biochemical Society, London.
Ogren, W. L. & Chollet, R. (1982) In: *Photosynthesis*, Vol. II (Govindjee, ed.) p. 191. Academic Press.
Ohtaki, E. & Matsui, M. (1982) *Bound. Layer Met. 24*, 109.
Okada, M., Kitajima, M. & Butler, W. L. (1976) *Plant Cell Physiol. 17*, 35.
Öquist, G. (1981) In: *Physiological processes limiting plant productivity* (C. B. Johnson, ed.) p. 53. Butterworths.
Öquist, G. (1983) In: *Effects of stress on photosynthesis* (R. Marcelle, H. Clijsters & M. van Poucke, eds.) p. 211. Junk.
Öquist, G., Brunes, L., Hällgren, J. E., Gezelius, K., Hallén, M. & Malmberg G. (1980) *Physiol. Plant., 44*, 300.
Ormrod, D. P. (1982) In: *Effects of gaseous air pollution in agriculture and horticulture* (M. H. Unsworth & D. P. Ormrod, eds.) p. 307. Butterworths.
Osmond, C. B. (1981) *Biochem. Biophys. Acta 639*, 77.
Osmond, C. B., Allaway, W. G., Sutton, B. G., Troughton, J. H. Quieroz, O., Lüttge, W. & Winter, K. (1973) *Nature 246*, 41.
Osmond, C. B. & Holtum, J. A. M. (1981) In: *The biochemistry of Plants Vol. 8* (Hatch, M. D., Boardman, N. K., eds.) p. 283. Academic Press.
Osmond, C. B., Winter, K. & Powles, S. B. (1980) In: *Adaptation of plants to water and high temperature stress* (N. C. Turner, P. J. Kramer, eds.) p. 139. Wiley-Interscience.
Osmond, C. B., Winter, K. & Ziegler, H. (1982) In: *Encycl. Plant Physiol. N. S. Vol. 12B*, (O. L. Lange, P. S. Nobel, C. B. Osmond, H. Ziegler, eds.) p. 479. Springer-Verlag.
Overdieck, D. (1979) *Ber Deutsch. Bot. Ges. 91*, 633.
Overrein, L. N., Seip, H. M. & Tollan, A. (1980) In: *Acid Precipitation – Effects on Forests and Fish*, p. 18. Norwegian National Research Programme, Oslo.
O'Toole, J. C., Ozbun, J. L. & Wallace, D. H. (1977) *Physiol. Plant. 40*, 111.
Owen, P. C. (1968) *Aust. J. Expl. Agric. 8*, 582.
Parkinson, K. J. & Day, W. (1983) In: *Effects of stress on photosynthesis* (R. Marcelle, H. Clijsters, M. van Poucke, eds.) p. 65. Junk.
Pate, J. S., Layzell, D. B. & Atkins, C. A. (1979) *Plant Physiol. 64*, 1083.
Patriquin, D. G. (1978) *Aquatic Bot. 4*, 193.
Patterson, D. T. & Flint, E. P. (1980) *Weed Sci. 28*, 71.
Pearcy, R. W. (1977) *Plant Physiol. 59*, 795.
Pearcy, R. W., Berry, J. A. & Fork, D. C. (1977) *Plant Physiol. 59*, 873.
Pearman, I. S., Thomas, S. M. & Thorne, G. N. (1979) *Ann. Bot. 43*, 613.
Peisker, M., Ticha, I. & Apel, A. (1979) *Biochem. Physiol. Pfl. 174*, 391.
Peleg, M. (1976) *Water Res. 10*, 361.
Pendleton, J. W., Smith, G. E., Winter, S. R. & Johnston, T. J. (1968) *Agron. J. 60*, 422.
Penning de Vries, F. W. T., Brunsting, A. H. M. & van Laar, H.-H. (1974) *J. Theor. Biol. 45*, 339.
Pepper, G. E., Pierce, R. B. & Mock, J. J. (1977) *Crop Sci. 17*, 883.
Perry, A. M. (1982) In: *Carbon Dioxide Reviews* (W. C. Clark, ed.) p. 337. Oxford University Press, New York.
Philips, A. (1980) *Ann. Rev. Plant Physiol. 31*, 29.
Pike, C. S. & Berry, J. A. (1979) *Carn. Inst. Wash. Year Book 79*, 163.
Pike, C. S., Berry, J. A. & Raison, J. K. In: *Low temperature stress in crop plants: the role of the membrane.* (J. M. Lyons, D. Graham, J. K. Raison, eds.) p. 305. Academic Press.
Pisek, A., Larcher, W., Moser, W. & Pack, J. (1969) *Flora 158*, 603.
Pitter, R. L. (1977) *Agric. Met. 18*, 115.

Plaut, Z., Platt, S. & Bassham, J. A. (1976) *Plant Physiol. 57,* Suppl. 58.
Pollack, J. B. (1982) In: *Climate in Earth History* (W. H. Berger & J. C. Cromwell, eds.) p. 68. National Academy Press, Washington, D.C.
Polojakoff-Mayber, A. & Gale, J. (1975) *Plants in saline environments.* 213 pp. Springer-Verlag.
Portis, A. E., Chon, C. J., Mosback, A. & Heldt, H. W. (1977) *Biochim. Biophys. Acta 461,* 313.
Possingham, J. V. (1970) *Recent advantages in plant nutrition Vol. 1* (R. M. Samish, ed.) p. 155. Gordon and Breach, New York.
Potter, J. R. & Boyer, J. S. (1973) *Plant Physiol. 51,* 993.
Potter, J. R. & Jones, J. W. (1977) *Plant Physiol. 59,* 10.
Powles, S. B., Berry, J. A. & Björkman, O. (1980) *Carn. Inst. Wash. Year Book 79,* 157.
Powles, S. B., Berry, J. A. & Björkman, O. (1983) *Plant Cell Env. 6,* 117.
Powles, S. B., Chapman, K. S. R. & Osmond, C. B. (1980) *Aust. J. Plant Physiol. 7,* 737.
Powles, S. B. & Critchley, C. (1980), *Plant Physiol. 65,* 1181.
Prasad, B. J. & Rao, D. N. (1979) *Acta Bot. Indica. 7,* 16.
Prioul, J.-L. (1982) In: *Trends in photobiology* (C. Heléne, M. Charlier, T. M. Montenay-Gareftier, G. Lauftriat, eds.) p. 633. Plenum Press.
Prioul, J.-L. & Chartier, P. (1977) *Ann. Bot. 41,* 789.
Puckett, K. J., Neibor, E., Flora, W. P. & Richardson, D. H. S. (1973) *New Phytol, 72,* 141.
Puckett, K. J., Richardson, D. H. S., Flora, W. P. & Niebor, E. (1974) *New Phytol. 73,* 1183.
Puckridge, D. W. (1969) *Aust. J. Agric. Res. 20,* 623.
Puckridge, D. W. & Ratkowsky, D. A. (1971) *Aust. J. Agric. Res. 22,* 11.
Radford, P. J. (1967) *Crop Sci. 7,* 171.
Raghavendra, A. S. & Das, V. S. R. (1978) *Photosynthetica 12,* 200.
Raison, J. K. (1980) In: *Biochemistry of Plants Vol. 4* (P. K. Stumpf, ed.) p. 57. Academic Press.
Raison, J. K., Berry, J. A., Armond, P. A. & Pike, C. S. (1980) In: *Adaptations of plants to water and high temperature stress* (N. C. Turner, P. Kramer, eds.) p. 261. Wiley-Interscience.
Raison, J. K., Chapman, E. A., Wright, L. C. & Jacobs, S. W. L. (1979) In: *Low temperature stress in crop plants: the role of the membrane.* (J. M. Lyons, D. Graham, J. K. Raison, eds.) p.177. Academic Press.
Raschke, K. (1975) *Ann. Rev. Plant Physiol. 26,* 309.
Rawson, H. M., Begg, J. E. & Woodward, R. G. (1977) *Planta 134,* 5.
Redmann, R. E. (1978) *Can. J. Bot. 56,* 1999.
Redshaw, A. J. & Meidner, H. (1972) *J. Exp. Bot. 23,* 229.
Reed, K. L., Hammerly, E. R., Dinger, B. E. & Jarvis, P. G. (1976) *J. Appl. Ecol. 13,* 925.
Reid, R. A. & Leech, R. M. (1980) *Structure and Function of Subcellular Organelles.* 176 pp. Blackie, Glasgow.
Reimer, T. O., Barghoorn, E. S. & Margulis, L. (1979) *Precamb. Res. 9,* 93.
Rhodes, I. (1972) *J. Agric. Sci. 78,* 509.
Rhodes, I. (1973) *Herb. Abstr. 43,* 129.
Rhodes, I. & Mee, S. S. (1979) *Grass For. Sci. 35,* 39.
Richards, F. J. (1959) *J. Exp. Bot. 10,* 290.
Ripley, E. A. & Redmann, R. E. (1976) In: *Vegetation and the atmosphere Vol. 2* (J. L. Monteith, ed.) p. 351. Academic Press.
Robberecht, R., Caldwell, M. M. & Billings, W. D. (1980) *Ecology 61,* 612.
Roberts, J. & Wareing, P. F. (1975) *Ann. Bot. 39,* 311.
Robinson, S. P. & Walker, D. A. (1981) In: *The biochemistry of plants Vol. 8* (Hatch, M. D., Boardman, N. K., eds.) p. 193. Academic Press.
Rodin, L. E. & Basilevic, N. I. (1966) *Production and mineral cycling of terrestrial vegetation* (English translation, G. E. Fogg, ed.). 253 pp. Oliver & Boyd.

REFERENCES

Rogers, H. H., Sinclair, T. R. & Heck, W. W. (1980) *Plant Physiol. 65 Suppl.*, 49.
Rook, D. A. (1969) *N.Z. J. Bot. 7*, 43.
Rook, D. A. & Corson, M. J. (1978) *Oecologia 36*, 371.
Ross, J. K. (1969) *Prediction and measurement of photosynthetic productivity.* (Setlik, I., ed.) p. 29. Wageningen.
Ross, J. K. (1975) *Rediaciomnyj rezim c archtektonika rastitel nogo pokroua* (Leningrad: Gidro meteoizdat) 342 pp. (in Russian.)
Ross, J. K. & Nilson, T. (1967) *Photosynthesis of production systems.* (Nichiporovich, A. A., ed.) 188 pp. Israel Programme for Scientific Translations, Jerusalem.
Rowley, J. A. & Taylor, A. O. (1972) *New Phytol. 71*, 447.
Russell, G. & Grace, J. (1978) *J. Exp. Bot. 29*, 245.
Ryle, G. J. A., Powell, C. E. & Gordon, J. A. (1979) *J. Exp. Bot. 30*, 135.
Sagisaki, S. (1974) *Kagcka to Seibutsu, 12* (in Japanese).
Sanchez-Diaz, M. F. & Kramer, P. J. (1971) *Plant Physiol. 48*, 613.
Sanderson, F. H. (1975) *Science 188*, 503.
Sane, P. V. (1977) In: *Encycl. Plant Physiol. N. S. Vol. 5* (A. Trebst, M. Avron, eds.) p. 522. Springer-Verlag.
Santarius, K. A. (1975) *J. Therm. Biol. 1*, 101.
Satoh, K. (1970) *Plant Cell Physiol. 11*, 15.
Satoh, K. & Fork, D. C. (1982) *Plant Physiol. 70*, 1004.
Saugier, B. (1970) *Oecol. Plant. 5*, 179.
Sawada, S., Metsushima, H. & Miyachi, S. (1974) *Plant Cell Physiol. 15*, 239.
Sawada, S. & Miyachi, S. (1974) *Plant Cell Physiol. 15*, 111.
Schantz, H. L. & Piemiesel, L. N. (1927) *J. Agric. Res. 34*, 1093.
Schlesinger, W. H. (1977) *Ann. Ecol. Syst. 8*, 51.
Schmidt, A. (1979) In: *Encycl. Plant Physiol. N. S. Vol. 6.* (M. Gibbs, E. Latzko, eds.) p. 481. Springer-Verlag.
Schnabl, H. & Ziegler, H. (1974) *Ber. Deutsch. Bot. Ges. 87*, 13.
Schreiber, U. & Berry, J. A. (1977) *Planta 136*, 233.
Schulze, E.-D., Fuchs, M. & Fuchs, M. I. (1977a) *Oecologia 30*, 329.
Schulze, E.-D., Fuchs, M. I. & Fuchs, M. (1977b) *Oecologia 29*, 43.
Schulze, E.-D. & Hall, A. E. (1981) In: *Physiological processes limiting plant productivity* (C. B. Johnson, ed.) p. 217. Butterworths.
Schulze, E.-D. & Koch, W. (1969) In: *Productivity of forest ecosystems.* p. 141, Unesco.
Schulze, E.-D., Lange, O. L. & Lemke, G. (1972) *Planta 10*, 15.
Seely, G. R. (1977) In: *The intact chloroplast. Vol. 2* (J. Barber, ed.) p. 1. Elsevier.
Seeman, J. R. & Berry, J. A. (1982) *Carnegie Inst. Wash. Yearbk. 81*, 78-83.
Sestak, Z., Catzky, J. & Jarvis, P. G. (1971) *Plant Photosynthetic Production – a manual of methods* 818 pp. Junk.
Shavit, N. (1980) *Ann. Rev. Biochem. 49*, 111.
Shearman, L. L., Eastin, J. D., Sullivan, C. Y. & Kinbader, E. J. (1972) *Crop Sci. 12*, 406.
Sheehy, J. E. (1977) *Ann. Bot. 41*, 593.
Sheehy, J. E. & Peacock, J. M. (1977) *Ann. Bot. 41*, 567.
Shibles, R. M. & Weber, C. R. (1965) *Crop Sci. 5*, 575.
Shibles, R. M. & Weber, C. R. (1966) *Crop Sci. 6*, 55.
Shimshi, D. (1969) *J. Exp. Bot. 20*, 381.
Shirahashi, K., Hiyakawa, S. & Sugiyama T. (1978) *Plant Physiol. 62*, 826.
Shouse, P., Dasberg, S. D., Jury, W. A. & Stolzy, L. H. (1977) In: *Proc. int. symp. on rainfed agriculture in semi-arid regions* p. 424. University of California, Riverside.
Sinclair, T. R., Allen, L. H. & Stewart, D. W. (1971) *Proc. 1970 Computer Simulation Conf. La Jolla, Cal.* p. 784.
Sinclair, T. R., Murphy, C. E. & Knoerr, K. R. (1976) *J. Appl. Ecol. 13*, 813.
Singh, J. S., Lauenroth, W. K. & Steinhorst, R. K. (1975) *Bot. Rev. 41*, 181.
Singh, J. S., Tolica, M. J., Risser, P. G., Redmann, R. E. & Marshall, J. K. (1980)

Grasslands, Systems Analysis and Man (Breymeyer, A. I. & van Dyne, G. N., eds.) p. 59. Cambridge University Press.
Sisson, W. B. & Caldwell, M. M. (1976) *Plant Physiol. 58,* 563.
Sisson, W. B. & Caldwell, M. M. (1977) *J. Exp. Bot. 28,* 691.
Slack, C. R., Rougham, P. G. & Basset, H. C. N. (1974) *Planta 118,* 57.
Slatyer, R. O. (1977a) *Aust. J. Plant Physiol. 4,* 301.
Slatyer, R. O. (1977b) *Aust. J. Plant Physiol. 4,* 901.
Slatyer, R. O. & Ferrar, P. J. (1977) *Aust. J. Plant Physiol. 4,* 595.
Slatyer, R. O. & Morrow, P. A. (1977) *Aust. J. Bot. 25,* 1.
Smith, A., Woolhouse, H. W. & Jones, D. A. (1982) *Planta 156,* 441.
Smith, B. N. (1972) *Bio. Science 22,* 226.
Smith, B. N. (1976) *Biosystems 8,* 24.
Smith, B. N. (1982) In: *CRC Handbook of biosolar resources Vol. 1 Part 2* (A. Mitsui, C. C. Black, eds.) p. 99. CRC Press, Boca Raton.
Smith, B. N. & Brown, W. V. (1973) *Amer. J. Bot. 60,* 505.
Smith, B. N. & Robbins, M. J. (1975) In: *Proc. 3rd Int. Phot. Cong.* (M. Avron, ed.) p. 1579. Elsevier.
Solov'ev, V. A. (1969a) *Fiziol. Rast. 16,* 498. (in Russian.)
Solov'ev, V. A. (1969b) *Fiziol. Rast. 16,* 870. (in Russian.)
Somerville, C. R. & Ogren, W. L. (1979) *Plant Physiol. 63,* Suppl. 152.
Spencer, D. & Possingham, J. V. (1960) *Aust. J. Biol. Sci. 13,* 441.
Spierings, F. H. F. G. (1971) *Neth. J. Plant Path, 77:* 194.
Srivastava, H. S., Jolliffe, P. A. & Runeckles, V. C. (1975) *Can. J. Bot. 53,* 466.
Stanhill, G., Fuchs, M., Bakker, J. & Moreshet, S. (1972) *Agric. Met. 11,* 385.
Stapleton, H. N., Buxton, D. R., Watson, F. L., Nolting, D. J. & Baker, D. N. (1973) *Tech. Bull. Ariz. Exp. Sta.* 206.
Steponkus, P. L. (1981) In: *Encycl. Plant Physiol. N. S. Vol. 12A* (O. L. Lange, P. S. Nobel, C. B. Osmond & H. Ziegler, eds.) p. 371. Springer-Verlag.
Stern, W. R. & Donald, C. M. (1962) *Aust. J. Agric. Res. 13,* 599.
Stewart, D. W. & Lemon, E. R. (1969) *U.S. Army Eng. Tech. Rep.* 2-68, 1.
Stewart, G. R., Larher, F., Ahmad, I. A. & Lee, J. A. (1979) In: *Ecological processes in coastal environments.* (R. L. Jeffries, A. L. Davy, eds.) p. 211, Blackwell.
Stowe, L. G. & Teeri, J. A. (1978) *Amer. Nat. 112,* 609.
Strogonov, B. P. (1962) *Physiological Basis of Salt Tolerance of Plants.* 279 pp. Israel Program for Scientific Translations, Jerusalem.
Stuiver, M. (1978) *Science, 199,* 253.
Sugiyama, T. & Boku, K. (1978) *Plant Cell. Physiol. 17,* 851.
Sugiyama, T., Schmitt, M. R., Ku, S. B. & Edwards, G. E. (1979) *Plant Cell Physiol. 20,* 965.
Sung, S. J. M. & Krieg, D. R. (1979) *Plant Physiol. 64,* 852.
Szabolc, I. (1974) *Salt Affected Soils in Europe,* 63pp. Martinus Nijhoff, The Hague.
Szabolc, I. (1979) *Review of research on salt affected soils.* Unesco, Paris.
Szaniawski, R. K. & Wierzbicki, B. (1978) *Photosynthetica 12,* 412.
Szarek, S. R. (1979) *Photosynthetica 13,* 467-473.
Szarek, S. R. & Ting, I. P. (1977) *Photosynthetica 11,* 330.
Tajima, K. (1971) *Proc. Crop Sci. Soc. Japan 40,* 247.
Takeda, I., Tajima, M. & Adki, M. (1976) *Proc. Crop Sci. Soc. Japan 45,* 139.
Tanaka, A., Kawano, K. & Yamaguchi, J. (1966) *Int. Rice Res. Inst. Tech. Bull. No. 7,* 46 pp. International Rice Research Institute, Laguna, Phillipines.
Tanaka, A., Matsushima, S., Kojyo, S. & Nitta, H. (1969) *Proc. Crop Sci. Soc. Japan 38,* 387.
Taniyama, T. (1972) *Bull. Fac. Agric. Mie. Univ., Japan 44,* 11.
Tanner, J. W., Gardner, C. J., Stoskoff, N. C. & Reinbergs, E. (1966) *Can. J. Plant Sci. 46,* 690.

REFERENCES

Taylor, A. O. & Rowley, J. A. (1971) *Plant Physiol. 47*, 713.
Taylor, A. O., Slack, C. R. & McPherson, H. G. (1974) *Plant Physiol. 54*, 696.
Taylor, S. E. & Sexton, O. J. (1972) *Ecology, 53*, 143.
Tazaki, T., Ishihara, K. & Ushijima, T. (1980) In: *Adaptation of plants to water and high temperature stress* (N. C. Turner, P. J. Kramer, eds.) p. 309. Wiley-Interscience.
Teeri, J. A. & Stowe, L. G. (1976) *Oecologia 23*, 1.
Teeri, J. A., Stowe, L. G. & Livingstone, D. A. (1980) *Oecologia 47*, 307.
Tenhunen, J. D., Weber, J. A., Filidek, L. H. & Gates, D. M. (1977) *Oecologia 30*, 189.
Teramura, A. H., Biggs, R. H. & Kossuth, S. V. (1980) *Plant Physiol. 65*, 483.
Terry, N. & Ulrich. A. (1973a) *Plant Physiol. 51*, 43.
Terry, N. & Ulrich. A. (1973b) *Plant Physiol. 51*, 783.
Tew, J., Cresswell, C. F. & Fair, P. (1975) *Proc. 3rd Int. Phot. Cong.* (M. Avron, ed.) p. 1249. Elsevier.
Thom, A. S. (1975) In: *Vegetation and the atmosphere Vol. 1* (J. Monteith, ed.) p. 7. Academic Press.
Thompson, L. M. (1969) *Agron, J. 61*, 453.
Thomson, W. W. (1975) *Responses of plants to air pollution* (J. B. Mudd & T. T. Koslowski, eds.) p. 179. Academic Press.
Thornber, J. P. (1975) *Ann. Rev. Plant Physiol. 26*, 127.
Thornber, J. P. & Alberte R. S. (1977) In: *Encycl. Plant. Physiuol. N. S. Vol. 5* (A. Trebst, M. Avron, eds.) p. 574. Springer-Verlag.
Thornley, J. H. M. (1976) *Mathematical Models in Plant Physiology* 318 pp. Academic Press.
Tieszen, L. L. & Sigurdson, D. C. (1973) *Arc. Alp. Res. 5*, 59.
Ting, I. P. & Gibbs, M. (eds.) (1982) *Crassulacean acid metabolism.* 316 pp. American Society of Plant Physiologists. Rockville, Maryland.
Tingey, G. L. & Taylor, G. E. (1982) In: *Effects of gaseous air pollution in agriculture and horticulture* (M. H. Unsworth & D. P. Ormrod, eds.) p. 113. Butterworths.
Tinus, R. W. (1974) *Agric. Met. 14*, 99.
Todd, G. W. & Basler, E. (1965) *Phyton. 22*, 79.
Tolbert, N. E. (1979) In: *Encycl. Plant Physiol. N. S. Vol. 6* (M. Gibbs, E. Latzko, eds.) p. 338. Springer-Verlag.
Tombesi, L., Cale, M. T. & Tiborne, B. (1969) *Plant Soil 31*, 65.
Tomlinson, G. H. & Silversides, C. R. (1982) *Acid Deposition and Forest Damage* Dorntar Inc., Canada.
Trebst, A. V. & Avron, M. eds. (1977) *Encycl. Plant Physiol. N. S. Vol. 5*, Springer-Verlag.
Tregunna, E. B., Smith, B. N., Berry, J. A. & Downton, W. J. S. (1970) *Can. J. Bot. 48*, 1209.
Treharne, K. J. & Eagles, C. F. (1970) *Photosynthetica 4*, 107.
Treharne, K. J. & Nelson, C. J. (1975) In: *Environmental and biological control of photosynthesis* (R. Marcelle, ed.) p. 61. Junk.
Trenbath, B. R. & Angus, J. F. (1975) *Field Crop Abstr. 28*, 231.
Tripathy, B. C., Bhatia, B. & Mohanty, P. (1981) *Biochim. Biophys. Acta. 638*, 217.
Troeng, E. & Linder, S. (1982a) *Physiol. Plant. 54*, 7.
Troeng, E. & Linder, S. (1982b) *Physiol. Plant. 54*, 15.
Troughton, J. H. (1969) *Aust. J. Biol. Sci. 22*, 289.
Troughton, J. H. (1971) *Photosynthesis & photorespiration* (M. D. Hatch, C. B. Osmond, eds.) p. 124. Wiley-Interscience.
Troughton, J. H. (1979) In: *Encycl. Plant Physiol. N. S. Vol. 6* (M. Gibbs, E. Latzko, eds.) p. 140. Springer-Verlag.
Troughton, J. H. & Slatyer, R. O. (1969) *Aust. J. Biol. Sci. 22*, 815.
Tschape, M. (1972) *Biochem. Physiol. Pfl. 163*, 81.
Tsunoda, S. (1962) *Jap. J. Breeding 12*, 49.
Turk, K. J. & Hall, A. E. (1980a) *Agron. J. 72*, 428.

Turk, K. J. & Hall, A. E. (1980b) *Agron. J. 72,* 434.
Turner, N. C., Begg, J. E. Rawson, M. M., English, S. D. & Hearn, A. B. (1978) *Aust. J. Plant Physiol. 5,* 179.
Turner, N. C. & Incoll, L. D. (1971) *J. Appl. Ecol. 8,* 581.
Turner, R. E. (1976) *Contrib. Mar. Sci. 20,* 47.
Uchijima, Z. (1976) *Vegetation and the atmosphere. Vol. 2.* (J. L. Monteith ed.) p. 33. Academic Press.
UK-ISES (1976) *Solar energy: a UK assessment.* Chap. 9. UK-ISES Publ., London.
Unsworth, M. H. (1981) In: *Physiological processes limiting plant productivity* (C. B. Johnson, ed.) p. 293. Butterworths.
Unsworth, M. H. (1982) In: *Effects of gaseous air pollution in agriculture and horticulture.* (M. H. Unsworth, D. P. Ormrod eds.) p. 43. Butterworths.
Unsworth, M. H. & Ormrod, D. P. (eds.) (1982) *Effects of gaseous air pollution in agriculture and horticulture* 532 pp. Butterworths.
Ursino, D. J., Hunter, D., Laing, R. Fowler, P. & Keighley, J. (1979) *Plant Physiol. 63,* Suppl. 12.
Usamanov, P. D., Abdullaev, K. A., Pinkhasov, Y. I. & Bikaslyari, G. R. (1975) *Genetika 11,* 22.
Van Assche, F. & Clijsters, H. (1983) In: *Effects of stress on photosynthesis* (R. Marcelle, H. Clijsters, M. van Poucke, eds.) p. 371. Junk.
Van Assche, F. Clijsters, H. & Marcelle R. (1979) In: *Photosynthesis and plant development* (R. Marcelle, H. Clijsters, M. van Poucke, eds.) p. 175. Junk.
Van Hasselt, Ph. R. (1972) *Acta. Bot. Neerl. 21,* 539.
Van Hasselt, Ph. R. & Strikverde, J. T. (1976) *Physiol. Plant. 37,* 253.
Van Hasselt, Ph. R. & van Bierlo, H. A. C. (1980) *Physiol. Plant. 50,* 52.
Van, T. K. & Garrard, L. A. (1976) *Soil Soc. Fla. Proc. 35,* 1.
Van, T. K., Garrard, L. A. & West, S. H. (1976) *Crop Sci. 16,* 715.
Varlet-Grancher, C., Bonhomme, R., Chartier, P. & Artis, P. (1982) *Acta Oecol. Plant. 3,* 3.
Velthuys, B. R.(1980) *Ann. Rev. Plant Physiol. 31,* 545.
Venus, J. C. & Causton, D. R. (1979) *Ann. Bot. 43.* 623.
Vergara, B. S. (1977) *Crop Physiology* (U.S. Gupta, ed.) p. 137. Oxford & IBH Publ., New Delhi.
Verhagen, A. M. W., Wilson, J. H. & Britten, E. J. (1963) *Ann. Bot. 27,* 641.
Vernon, A. J. & Allison, J. C. (1963) *Nature 200,* 814.
Vogel, J. C. (1980) *Fractionation of carbon isotopes during photosynthesis.* 135 pp. Springer-Verlag.
de Vries, C. A., Ferwerda, J. D. & Flach, M. (1967) *Neth. J. Agric. Sci. 15,* 241.
Vu, C. V. Allen, R. H. & Garrard, L. A. (1981) *Physiol. Plant. 52,* 353.
Waggoner, P. E. (1969) *Crop Sci. 9,* 315.
Walker, D. A. (1981) In: *Proc. 5th Int. Phot. Cong. Vol. 4* (G. Akoyunoglou, ed.) p. 189. Balaban International, Philadelphia.
Walker, D. A. & Robinson, S. P. (1978) *Ber. Deutsch. Bot. Ges. 91,* 513.
Warren-Wilson, J (1965) *J. Appl. Ecol. 2,* 383.
Warren-Wilson, J. (1972) In: *Crop processes in controlled environments.* (A. R. Rees, K. E. Cockshull, D. W. Hand & R. G. Hurd, eds.) p. 7 Academic Press.
Warrington, I. J., Dixon, T., Robotham, R. W. & Rook, D. A. (1978b) *J. Agric. Eng. Res. 23,* 23.
Warrington, I. J., Edge, E. A., & Green, L. M. (1978a) *Ann. Bot. 42,* 1305.
Warrington, I. J. & Mitchell, K. J. (1975) *J. Agric. Eng. Res. 20,* 295.
Warrington, I. J. & Mitchell, K. J. (1976) *Agric. Met. 16,* 247.
Warrington, J. J., Mitchell, K. J. & Halligan, G. (1976) *Agric. Met. 16,* 231.
Watson, D. J. (1947) *Ann. Bot. 11,* 42.
Watson, D. J. (1952) *Adv. Agron. 4,* 101.

REFERENCES

Watson, D. J. (1958) *Ann. Bot. 22,* 37.
Watson, D. J. (1971) In: *Potential crop production* (P. F. Wareing & J. P. Cooper, eds.) p. 76. Heinemann.
Watson, D. J. & Saw, S. A. W. (1962) *Ann. Appl. Biol. 50,* 1.
Watt, G. D., Bulen, W. A., Burns, A. & Hadfield, K. A. (1975) *Biochemistry 14,* 4266.
Watts, W. R., Neilson, R. E. & Jarvis, P. G. (1976) *J. Appl. Ecol. 8,* 925.
Webb, W., Szasrek, S. Lavenroth, W., Kinerson, R. & Smith, M. (1978) *Ecology, 59,* 1239.
Weinstein, L. H. & McCune, D. C. (1979) *Stress physiology in crop plants* (M. Mussel & R. C. Staples, eds.) p. 327. Wiley.
Wellburn, A. R. (1982) In: *Effects of gaseous air pollution in agriculture and horticulture* (M. H. Unsworth & D. P. Ormrod, eds.) p. 169. Butterworths.
Wellburn, A. R., Majernik, O. & Wellburn, F. A. M. (1972) *Env. Poll. 3,* 37.
Wellman, E. (1974) *Ber. Deutsch. Bot. Ges. 87,* 267.
Whittaker, R. H. & Woodwell, G. M. (1968) *J. Ecol. 56,* 1.
Whittaker, R. H. & Likens, G. E. (1975) *Ecol. Stud. 14,* 305.
Whittingham, C. P. (1981) In: *Proc. 5th Int. Phot. Cong. Vol. 6* (G. Akoyuroglou, ed.) p. 3. Balaban International, Philadelphia.
Wiegert, R. G. & Evans, F. C. (1964) *Ecology 45,* 49.
Wielgolaski, F. E. (1977) In: *Proc. 4th Int. Phot. Cong.* (D. O. Hall, J. Coombs & T. W. Goodwin, eds.) p. 245. Biochem. Soc., London.
Wifong, R. T., Brown, R. H. & Blaser, R. E. (1967) *Crop Sci. 7,* 27.
Wilhelm, W. W. & Nelson, C. J. (1979) *Crop Sci. 18,* 951.
Williams, C. J. & Kemp, P. R. (1978) *Bot. Gaz. 139,* 150.
Williams, W. A., Loomis, R. S. & Lepley, C. R. (1965) *Crop Sci. 5,* 211.
Wilson, D. (1972) *J. Exp. Bot. 23,* 517.
Wilson, D. & Cooper, J. P. (1967) *Nature 214,* 989.
Wilson, D., Eagles, C. F. & Rhodes, I. (1980) In: *Opportunities for increasing crop yields* (R. G. Hurd, P. V. Biscoe, C. Dennis, eds.) p. 21. Pitman.
Wilson, J. M. (1979) In: *Low temperature stress in crop plants: the role of the membrane* (J. H. Lyons, D. Graham, J. K. Raison, eds.) p. 47. Academic Press.
Wilson, V. E. & Teare, I. D. (1972) *Crop Sci. 12,* 507.
Winner, W. E. & Mooney, H. A. (1980a) *Oecologia 44,* 290.
Winner, W. E. & Mooney, H. A. (1980b) *Oecologia 44,* 296.
Winner, W. E. & Mooney, H. A. (1980c) *Oecologia 46,* 49.
Winter, K. (1974) *Plant Sci. Lett. 3,* 379.
Winter, K. Lüttge, U., Winter, E. & Troughton, J. H. (1978) *Oecologia 34:* 225.
Winter, S. R. & Ohlrogge, A. J. (1973) *Agron, J. 65,* 395.
de Wit, C. T. (1965) *Agric. Res. Rep. 663* 57 pp. Wageningen.
de Wit, C. T. (1978) *Simulation of Assimilation, Respiration and Transpiration of Crops* 141 pp. Wageningen.
Wittwer, S. H. (1975) *Science 188,* 579.
Wittwer, S. H. (1977) *Crop Physiology* (U. S. Gupta ed.) p. 310. Oxford IBH, New Delhi.
Wittwer, S. H. (1980a) *The biology of crop productivity* (P. S. Carlson, ed.) p. 413. Academic Press.
Wittwer, S. H. (1980b) *Paper for CIMMYT Long Range Planning Conf.* Centro Internacional de Majoramiento de Maiz y Trigo, México D.F.
Wittwer, S. H. (1982) *New Scientist 95 (1315),* 233.
Wolfe, J. (1978) *Plant Cell Env. 1,* 241.
Wong, S. C. (1979a) *Oecologia 44,* 68.
Wong, S. C. (1979b) *Nature, 282,* 424.
Wong, S. C., Cowan, I. R. & Farquhar, G. D. (1978) *Plant Physiol. 62,* 670.
Woolhouse, H. W. (1968) *Hilger, J. 13,* 7.
Woolhouse, H. W. (1978) *Endeavour 2,* 35.
Wyn Jones, R. G. (1981) In: *Physiological processes limiting plant productivity* (C. B.

Johnson, ed.) p. 271. Butterworths.
Wyn Jones, R. G., Storey, R., Leigh, R. A., Ahmad, N. & Pollard, A. (1977) In: *Regulation of cell membrane activities in plants* (E. Marré, O. Ciferri eds.) p. 121. Elsevier.
Yocum, C. S., Allen, L. H. & Lemon, E. R. (1964) *Agron. J. 56,* 249.
Yordanov, J. T. & Vasileva, U.S. (1976) *Fiziol. Rast. 23,* 812. (in Russian.)
Yoshida, S. (1972) *Ann. Rev. Plant Physiol. 23,* 437.
Yu, S., Liu, Y., Li, Z., Yang, W., Wu, Y. (1982) In: *Effects of gaseous air pollution in agriculture and horticulture* (M. H. Unsworth & D. P. Ormrod, eds.) p. 507. Butterworths.
Zelitch, I. (1973) *Proc. Natl. Acad. Sci. 70,* 579.
Zelitch, I. (1979) In: *Encycl. Plant Physiol. N. S. Vol. 6* (M. Gibbs & E. Latzko, eds.) p. 353. Springer-Verlag.
Zelitch, I. (1982) *Bio. Sci. 32,* 796.
Zelitch, I. & Day, P. R. (1973) *Plant Physiol. 52,* 33.
Zewlawski, W., Szaniawski, R., Dybczynski, W. & Piechurowski, A. (1973) *Photosynthetica 7,* 351.
Ziegler, I. (1972) *Planta 103,* 155.
Ziegler, I. (1973) *Phytochemistry 12,* 1027.
Ziegler, I. (1974) *Phytochemistry 13,* 2403.
Ziegler, I. (1975) *Residue Rev. 56,* 79.
Zelitch, I. (1982) *Bio Sci. 32,* 796.